Fay, Peter.
    The blue-greens : (cyanophyta-
cyanobacteria) / Peter Fay. -- London ;
Baltimore, Md., U.S.A. : E. Arnold,
1983.
    88 p. : ill. ; 22 cm. -- (The
Institute of Biology's studies in
biology, ISSN 0537-9024 ; no. 160)
    Bibliography: p. [85]
    Includes index.
    ISBN 0-7131-2878-X (pbk.)

1. Cyanobacteria.   I. Title

The Institute of Biology's
Studies in Biology no. 160

The Blue-Greens

(Cyanophyta – Cyanobacteria)

Peter Fay

Professor of Botany

Westfield College
University of London

Edward Arnold

© Peter Fay 1983

First published in Great Britain 1983
by Edward Arnold (Publishers) Ltd
41 Bedford Square, London WC1 3DQ

Edward Arnold (Australia) Pty Ltd
80 Waverly Road
Caulfield East 3145
PO Box 234
Melbourne

Edward Arnold
300 North Charles Street
Baltimore
Maryland 21201
U.S.A.

British Library Cataloguing in Publication Data

Fay, Peter
    The blue-greens (Cyanophyta-Cyanobacteria).—
    (Studies in biology/Institute of Biology; no. 160)
    1. Blue-green algae
    I. Title   II. Series
    589.4'6   QK569.C96

    ISBN 0-7131-2878-X

Printed and bound in Great Britain at
The Camelot Press Ltd, Southampton

General Preface to the Series

Because it is no longer possible for one textbook to cover the whole field of biology while remaining sufficiently up to date, the Institute of Biology proposed this series so that teachers and students can learn about significant developments. The enthusiastic acceptance of 'Studies in Biology' shows that the books are providing authoritative views of biological topics.

The features of the series include the attention given to methods, the selected list of books for further reading and, wherever possible, suggestions for practical work.

Readers' comments will be welcomed by the Institute.

1983   Institute of Biology
20 Queensberry Place
London SW7 2DZ

Preface

This book is intended as a brief introduction to the study of the cyanobacteria of blue-green algae, a group of micro-organisms considered to be highly developed bacteria and/or very simple plants. In fact, they represent a very specific lineage in the biological evolution and display physiological capabilities which in many respects surpass those found among bacteria and higher plants. I have attempted to illustrate the many interesting aspects of their evolutionary history, their structure and function, and their ecological and economic importance. I hope that the book will not only explain why blue-greens are increasingly attracting the attention of research workers from diverse disciplines of pure and applied biology but that it may also provide a stimulus to further study of these unique organisms.

My thanks are due to my colleagues, particularly to Professor A.D. Boney, Professor A.E. Walsby and Mr Michael Wyman, for their helpful comments on the manuscript, to those, to whom acknowledgement by name is made where appropriate, who have generously provided illustrations, and to all other authors and publishers for permission to reproduce previously published information. I am also grateful to Mr Y. Sahota for help in photographic work and to my publishers for their help and advice in preparing the manuscript for the press.

London, 1983   P.F.

Contents

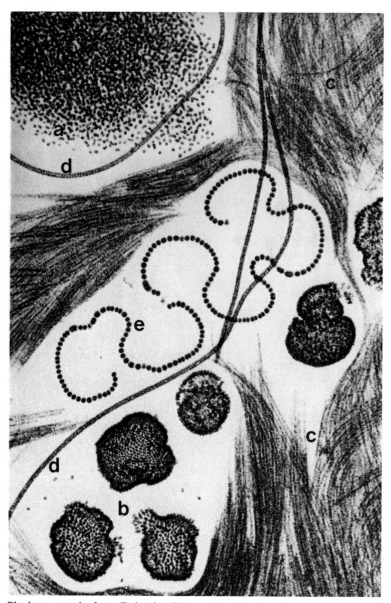

Plankton sample from Esthwaite Water, Lake District, showing colonies of (**a**) *Microcystis* and (**b**) *Coelosphaerium,* (**c**) filament bundles of *Aphanizomenon,* and fiaments of (**d**) *Oscillatoria* and (**e**) *Anabaena* species. By courtesy of Dr H.M. Canter.

Introduction

Blue-green algae or Cyanobacteria are a remarkable group of simple photosynthetic micro-organisms. In evolutionary terms they represent a link between bacteria and green plants. Their cellular organization, known as prokaryotic, is characterized by the lack of membrane-bound organelles such as a true nucleus, a chloroplast or a mitochondrion, and resembles that found in bacteria. Hence the genetic material, the photosythetic apparatus and the respiratory system are not segregated by means of internal membranes from the rest of the cell. Their principal mode of nutrition, oxygen-evolving photosynthesis, however, is similar to that which operates in all other nucleate or eukaryotic algae and in green plants.

Blue-greens apparently have a long history on Earth. Their fossils were recently identified in sediments from the Early Precambrian period, over three billion years old. At that time they were probably the chief primary producers of organic matter and the first organisms to release elemental oxygen, O_2, into the primitive atmosphere, which was until then free from O_2. Thus blue-greens were most probably responsible for a major evolutionary transformation leading to the development of aerobic metabolism and to the subsequent rise of higher plant and animal forms.

In spite of being an ancient group, blue-greens are nevertheless widespread and often abundant in the contemporary global scene. Although they are minute organisms, their mass occurrence makes them clearly visible to naked eye. There are many commonly known examples of their presence in nature. Floating filaments and spherical colonies of blue-greens are familiar components of the microscopic community of freshwater and marine phytoplankton. They may occasionally occur in such abundance as to cause a striking green colouration to natural waters, giving them the appearance of a pea soup. Such 'water blooms' are frequent during the summer in the English lakes and in other temperate standing waters, and may persist almost permanently in tropical lakes. They attract even more attention when they accumulate on a calm summers day at the surface of lakes and water reservoirs forming a thick slimy scum which resembles spilt green paint. Such less pleasing occurrences of blue-greens are remembered by their repulsive appearance and the accompanying pungent odour.

Blue-greens are also common in the intertidal zone of marine shores, on damp rocks, salt marshes, on the beds of rivers, or on trunks of trees and stones. They form gelatinous masses, furry cushions, skin-like sheets or powdery coverings on the substratum. The colour of these forms may vary from blackish green to olive-green, orange-yellow or reddish brown, besides the typical blue-green. Blue-greens are found on moist soil and on wet peat.

The terrestrial *Nostoc commune* may appear in the form of olive-green gelatinous ball-like colonies often reaching the size of a walnut. Thermophilic (heat-tolerant or heat dependent) blue-greens aggregated in brilliant mats are the dominant organisms of hot springs. Blue-greens are also present in the more artificial urban environment forming dark green coatings on moist walls, on water tanks or on the surface of flower pots. They are a disappointing sight to the aquarist as they spread over the glass walls of a household aquarium, and when they grow over the surface of sand and water plants indicating that the biological equilibrium of the aquarium has been disturbed.

Blue-greens are of considerable importance in the natural environment as initial colonizers of arid land and as primary producers of organic matter. Perhaps of greater importance is their role in the fundamental process of biological nitrogen fixation. Many of the free-living blue-greens and those engaged in symbiotic associations with other plants and animals contribute significantly to the nitrogen fertility of aquatic and terrestrial habitats, including cultivated lands, particularly in the tropics. Their usually high content of protein makes them prospective candidates for new unconventional sources of animal and human food. Indeed, in the Central African state of Chad dried cakes of *Spirulina*, which grows abundantly in soda lakes, constitute an important part of the diet of local people. In the recent search for new sources of energy, some researchers consider blue-greens as the most promisng agents for the development of a biological solar energy conversion (or biophotolysis) system, based on the concurrent light-driven generation of elemental O_2 and H_2 through the action of blue-greens.

The study of blue-greens became 'fashionable' in the last ten years, and there are many academic and practical reasons for the recent upsurge of interest in them. First is the recognition of their important role in the natural environment. Another is the growing appreciation of the part blue-greens might play in Man's economy, particularly in solving the world's desperately urgent food and energy problems. A third and not least important reason for the rapidly-growing popularity of blue-greens is the realization that they provide a relatively simple and useful model system for the study of fundamental cellular processes such as macromolecular synthesis, regulation of gene expression, cell differentiation and ordered developmental pattern formation.

Finally, the controversy of naming blue-greens 'algae' or 'bacteria' warrant an explanation. Biological nomenclature as well as taxonomy and classification are abstract concepts created by scientists to facilitate our understanding and orientation in the extremely complex and diverse world of living organisms. Distinctions like species, genus or other groupings are either unclear or non-existent in nature. Modern taxonomy aims to reflect the basic similarities and phylogenetic relationships between various organisms. Classification requires a great deal of information and may become very complex with organisms which represent a transitionary stage in the evolutionary pathway. Is a virus an ordered assembly of macromolecules or a living organism? Is *Euglena* a plant or animal? These are well known examples of the inadequacy of taxonomic definitions. If one considers cellular organization or for example cell wall chemistry to be the most important characteristics of blue-greens, then they are justly called bacteria. There is no doubt that they belong to the Prokaryota. But if the similarities between blue-greens and chloroplasts, their principal mode of metabolism, or their contribution to the natural nutritional cycles are the prime matters of consideration, then blue-greens are rightly called, as before, algae. Clearly, *in sensu stricto* blue-greens are neither typical bacteria nor typical algae, and ignoring this fact can lead to unnecessary misconceptions. As for this book, we shall simply call them 'blue-greens'. Those who may wish to consult the literature for more details on any subject only briefly touched upon here should remember that the many names now in circulation (Cyanophyta, Myxophyta, Cyanochloronta, Cyanobacteria, blue-green algae, blue-green bacteria) refer to one and the same group of micro-organisms.

1 Form and Structure

Blue-greens are microscopic organisms and their structure cannot be studied without the considerable magnification provided by the light and electron microscopes. When examining a drop of the greenish lake water during an early summer 'bloom' under the microscope, a variety of micro-organisms may be seen in the water sample including algae, bacteria and fungi. Blue-greens, however, predominate in such waters, and can be recognized with a little practice on the basis of their pigmentation and the relatively homogenous appearance of their cytoplasm. The principal types of vegetative structure, namely the unicellular, colonial, simple filamentous and branched filamentous forms may easily be distinguished.

1.1 The range of typical forms and the major groups

Table 1 is a summary of the main characteristics of the major groups of blue-greens, and compares the early and more recent schemes of classification.

The simplest blue-greens appear to be the unicellular forms, most of which were placed in the order Chroococcales. Their cells are spherical, ovoid or cylindrical, and they all reproduce by binary fission. They occur singly if the daughter cells separate after cell division, as in the case of *Synechococcus* (Fig. 1–1a), or they may aggregate into irregular loose colonies, like those of *Microcystis* (Fig. 1–1b), which are held together by a slimy matrix secreted by the cells of the growing colony. Some genera, like *Merismopedia* or *Gloeocapsa*, produce more ordered colonies by means of a more or less regular sequence of cell division combined with sheath secretion (Fig. 1–1 c and d).

Another assemblage of unicellular; colonial blue-greens is distinguished by the particular mode of reproduction which may (or may not) supplement binary fission. In the order Chamaesiphonales this takes the form of budding (once referred to as exospore production) (Fig. 1–1e). In the order Dermocarpales and Pleurocarpales the principal mode of replication is by means of minute reproductive cells called baeocytes (earlier termed as endospores), produced as a result of multiple fission. This is essentially a series of rapid successive binary fissions which convert a single mother cell into many small daughter cells (Fig. 1–1f).

A great number of blue-green taxa display a filamentous morphology formed by the daughter cells remaining in close contact following repeated cell divisions which occur in a single plane at right angles to the main axis of the filament. This process gives rise to a simple multicellular structure which consists of a chain of cells and is called a trichome. The trichome may be straight, like that of *Oscillatoria* (Fig. 1–2a), or is coiled in a regular spiral as in

Table 1 The principal groups of the blue-greens (comparison of modern and early systems of classification).

Section[1]	Basic morphology	Reproduction	Plane of division	Order[2] (Family)	Representative genera
I	Unicellular or colonial	Binary fission	Single	Chroococcales	*Gloeobacter Gloeothece, Synechococcus, (Anacystis, Agmenellum) Gloeocapsa, Chroococcus, Synechocystis, Microcystis, Merismopedia*
II	Unicellular or colonial	Budding Multiple fission	Two or more	Chamaesiphonales Pleurocapsales	*Chamaesiphon Dermocarpa, Dermocarpella, Chroococcidiopsis Xenococcus, Myxosarcina, Pleurocapsa, Hyella*
III	Filamentous, non-differentiated	Trichome fragmentation, hormogonia	Single	Nostocales (Oscillatoriaceae)	*Oscillatoria, Microcoleus, Spirulina, Pseudanabaena, Plectonema, Lyngbya, Phormidium, Schizothrix*
IV	Filamentous, heterocystous	Trichome fragmentation, hormogonia, akinetes	Single	(Nostocaceae) (Rivulariaceae) (Scytonemataceae)	*Anabaena, Aphanizomenon, Nostoc, Nodularia, Anabaenopsis, Cylindrospermum Calothrix, Dichothrix, Gloeotrichia, Rivularia Scytonema, Tolypothrix*
V	Branched filamentous, heterocystous	Trichome fragmentation, hormogonia, akinetes	Two or more	Stigonematales	*Mastigocoleus, Nostochopsis, Mastigocladus, Westiella, Fischerella, Hapalosiphon, Stigonema, Chlorogloeopsis (Chlorogloea)*

[1]Rippka *et al.* (1979). *J. gen. Microbiol.*, **111**, 1–61.
[2]Fritsch (1945). *The Structure and Reproduction of Algae*, Vol 2. Cambridge University Press.

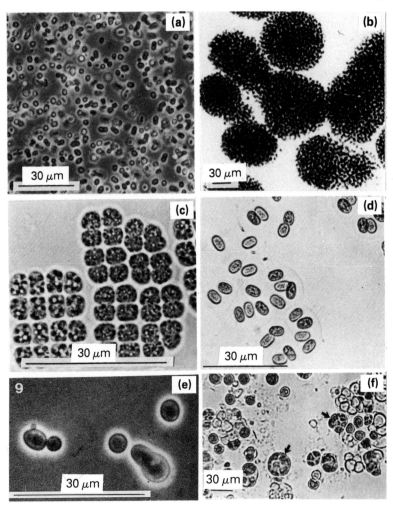

Fig. 1-1 Light micrographs of unicellular and colonial forms: (**a**) *Synechococcus*, (**b**) *Microcystis* colonies, (**c**) sheets of *Merismopedia*, (**d**) colonies of *Gloeothece*, (**e**) budding *Chamaesiphon*, (**f**) *Dermocarpa* producing baeocytes (arrows). By courtesy of Dr H.M. Canter (b,c). Dr R. Rippka (e) and Mr Z.S. Al-Ali (d).

Spirulina (Fig. 1–2b). Such variations in trichome morphology may occasionally appear within one and the same species. Cell size and shape also show great variability among the filamentous forms. In certain genera (e.g. *Oscillatoria*) the cylindrical or disc-shaped cells are closely appressed, the cross walls separating the cells are thin and scarcely visible under the light microscope, and the trichome is of uniform width throughout, appearing like a rod (Fig. 1–2a). In other groups (e.g. *Pseudanabaena*) distinct and often deep constrictions separate the neighbouring cells, thus limiting their contact to a narrow septum and conferring a necklace-like (moniliform) appearance to the trichome. Frequently the trichome is covered by a mucilaginous sheath (or envelope) which may appear diffluent or firm, and occasionally layered. Sheath formation seems to be induced and influenced by environmental factors which may be absent in culture. Trichome fragmentation within the sheath, and subsequent protrusion of one or both trichome fragments through the sheath, leads to the deceptive impression of a branched morphology (false branching) (Fig. 1–2c).

In the family Oscillatoriaceae the trichome is composed of cells essentially identical in appearance. But in other filamentous groups (Nostocaceae, Rivulariaceae, Scytonemataceae, and in the Stigonematales), cellular composition of the trichome is often heterogenous: cells differing in size and appearance may be seen interspersed between the normal (vegetative) cells of the trichome. These are of two kinds, heterocysts and akinetes (Fig. 2–4a).

Heterocysts are in general produced when combined forms of nitrogen, like ammonium–nitrogen or nitrate–nitrogen, are in short supply or absent from the growth medium. They are cells specialized for the prime function of nitrogen fixation (see Chapters 2 and 3). They are easily recognized by their usually rounded shape, pale colour, a distinct envelope, refractive structures ('polar nodules') near their attachment to the vegetative cells and the lack of cytoplasmic granulation. Akinetes are usually greatly enlarged cells and may be spherical, oval or cigar-shaped; they are surrounded by a massive envelope (capsule) and appear strongly granulated. Akinetes form under circumstances unfavourable for vegetative growth and function like spores, eventually giving rise to a new trichome (see § 2.5). Both heterocysts and akinetes are produced through the transformation (differentiation) of vegetative cells. With respect to their position within the trichome, heterocysts may be intercalary (interspersed between vegetative cells) or terminal; the former display two refractive nodules at each pole, the latter only at the single point of attachment to the vegetative cell. The relative positions of heterocysts and akinetes along the trichome appear to be genetically determined and give rise to a characteristic pattern of cell differentiation (see Chapter 2).

Trichomes are usually uniform in width but in the Rivulariaceae they may be tapered with a terminal heterocyst at the wider (basal) end of the trichome. Such tapered trichomes often terminate in thin colourless hairs which are apparently produced in response to nutrient deficiency (a shortage of a utilizable source of combined nitrogen, phosphate, iron or magnesium) (Fig. 1–2e).

In the filamentous forms discussed so far, the trichome is usually uniseriate and unbranched. A more complex situation develops in the order Stigonematales, partly by the formation of true lateral branches, and further by the main branch (primary trichome) becoming multiseriate. In addition, cells in the main branch and in the side branches may differ in size and shape (Fig. 1–2f).

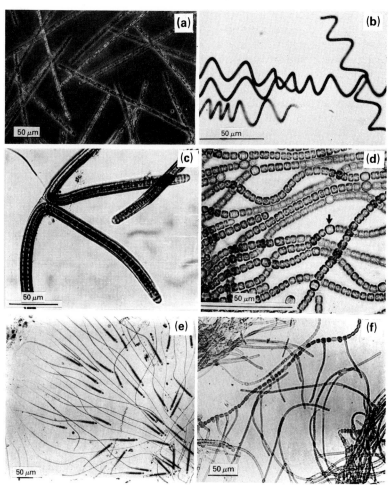

Fig. 1–2 Light micrographs of filamentous forms: (**a**) *Oscillatoria*, (**b**) *Spirulina*, (**c**) *Scytonema* showing false branching, (**d**) *Anabaena* with heterocysts (arrow), (**e**) *Gloeotricha* displaying basal heterocysts, akinetes and 'hairs', (**f**) *Mastigocladus* showing true branching. By courtesy of Prof. G.E. Fogg (c,e,f) and Mr W.E. Scott (d).

1.2 Cellular organization

The cells of most blue-greens measure from 2 to 5 μm (1 μm = 0.001 mm) in diameter. The detailed structure of such minute cells can be revealed only by the use of an electron microscope, which provides a magnification of between 10 000 and 250 000 times, sufficient to resolve the basic pattern of cellular construction. As mentioned earlier, the cells of blue-greens show a typical prokaryotic organization, profoundly different from the more complex eukaryotic architecture of plant and animal cells. It is nevertheless possible to make a rough distinction between two main regions of the cytoplasm; the peripheral and the central regions. The central (or nucleoplasmic) region is lightly granulated due to the presence of ribosomes, and contains the extremely folded thin thread of the prokaryotic type of 'circular' chromosome. The peripheral (or chromatoplasmic) region is traversed by thin sheets of the photosynthetic membrane system, incorporating the photosynthetic pigments. The cell is enclosed by a membrane, the plasmalemma, which is fortified by a multilayered cell wall. External to the wall, the cell may be surrounded by a gelatinous sheath or a more firm envelope (Fig. 1–6a).

1.3 Nuclear apparatus

The central nucleoplasmic region can be identified by its low electron density, when viewed in the electron microscope. The DNA (deoxyribonucleic acid) fibrils are organized in a complex helical and folded configuration and are distributed uniformly throughout the centroplasm. The DNA is probably associated with histone-like DNA-binding proteins and possibly with RNA (ribonucleic acid).

The size of the genome (the basic chromosome unit) varies widely in blue-greens, having a molecular weight between 1.6×10^9 and 8.6×10^9 daltons. Most unicellular forms possess genomes of about 1.6×10^9 to 2.7×10^9 daltons, similar in size to that found in other prokaryotic micro-organisms. Cells of the pleurocapsalean and filamentous blue-greens, however, contain larger genomes. The distribution of genome sizes among blue-greens is discontinuous and falls in four main categories with averages of 2.2, 3.6, 5.0 and 7.4×10^9 daltons. In general, strains with a more complex morphology possess larger genomes than those with a simpler vegetative structure. There is a theory that modern day prokaryotes evolved from a primitive bacterium with a small genome size of about 1.2×10^9 daltons. Replication of this genome seems to have occurred during biological evolution so that modern bacteria possess simple integer multiples of the primitive genome. The discontinuity of genome size categories and the correlation between genome size and morphological complexity support the hypothesis that genome evolution among blue-greens occurred by means of repeated duplication of the ancestral 1.2×10^9 daltons genome unit.

Ribosomes in blue-greens are seen diffusely distributed throughout the cytoplasm and particularly concentrated in the nucleoplasmic region. Their size, about 10 to 15 nm in diameter, and their sedimentation properties, 70 S (Svedberg units), are typically prokaryotic. The ribosomes dissociate into large 50 S and small 30 S subunits, and contain 5 S, 16 S and 23 S ribsomal RNAs with molecular weights of about 0.04, 0.55 and 1.05×10^6 daltons, respectively.

1.4 Photosynthetic membrane system

The peripheral region of the cytoplasm contains the photosynthetic apparatus of blue-greens. In unicellular forms this commonly consists of a few membrane layers which extend in concentric sheets beneath the cell membrane and surround the central nucleoplasmic region. The more elaborate membrane structure seen in the cells of filamentous species is apparently formed by means of expansion and invagination of the membranes. The basic structure of the membranes resembles a flattened sac and is known as the thylakoid. The appressed membranes of the thylakoid may occasionally or partially separate displaying the inner cavity of the thylakoid sac.

The lipid bilayer of the thylakoid membrane incorporates the lipophilic photosynthetic pigments, chlorophyll a and the various carotenoids (see Table 2 and Fig. 1–3). It also incorporates the components of the electron transport chain (like cytochromes, plastocyanin and ferredoxin). Chlorophyll a, which is the only chlorophyll species in blue-greens, is present in three different forms which can be distinguished on the basis of their absorption characteristics (maximum light absorption, λ_{max} at 670, 680 and 700 nm, respectively). Among the carotenoids, which absorb in the range between 480 and 520 nm wavelength, β-carotene appears to be universally present in all blue-greens while the presence and abundance of different xanthophylls (echinenone, zeaxanthin, oscillaxanthin, myxoxanthophyll) varies according to species.

An important part of the photosynthetic pigment complement of blue-greens is located in granular supramolecular complexes, called phycobilisomes. They contain the chromophore-bearing water-soluble phycobiliproteins, i.e. the blue phycocyanin (λ_{max} 620 nm) and allophycocyanin (λ_{max} 650–670 nm) and the red phycoerythrins (λ_{max} 550–570 nm). The relative quantities of these pigments may vary according to species and the spectral composition of light, and will determine the coloured appearance of these organisms. Phyco-

Table 2 The major photosynthetic pigments of blue-greens.

Group	Class	Major absorption bands (nm)
Chlorophylls	chlorophyll a	435, (670), 680, (700)
Carotenoids	β-carotene	(431), 450–454, 478–480
	echinenone	455–459, (475)
	zeaxanthin	(430), 453, 479
	canthaxanthin	466
	myxoxanthophyll	504–508, 474–476, 450–452
Phycobiliproteins	allophycocyanin	650
	phycocyanin	610–625
	phycoerythrin	555–565

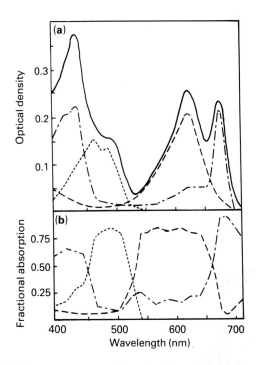

Fig. 1–3 (a) Absorption spectra of the cells of *Anacystis nidulans* and their pigment components. (b) Fraction of the total absorbed light absorbed by the component pigments at each wavelength. —————, cells; — · — · —, chlorophyll; – – – – –, phycobiliproteins; - - - - - -, carotenoids. (After Jones and Myers, 1964, *Plant Physiol.*, **39**, 938–46).

biliproteins represent the primary light-harvesting ('antenna') pigments for Photosystem II in blue-greens (see § 3.1). Light energy trapped by these pigments is transferred to chlorophyll with high efficiency. Each phycobilisome consists of a triangular core surrounded by six peripheral rods, composed of disc-shaped subunits (Fig. 1–4). The triangular core, which is in direct contact with the thylakoid membrane, consists of allophycocyanin while the rods contain phycocyanin and phycoerythrin. Phycobilisomes are arranged in parellel rows on the external surface of the thylakoid membrane. The close proximity to the thylakoid membrane is essential for maximum energy transfer by resonance in the primary act of photosynthesis.

1.5 Complementary chromatic adaptation

Environmental conditions are known to affect pigment composition of photosynthetic organisms. The supply of certain mineral nutrients (like nitrogen, phosphorus and iron) is essential for pigment synthesis. Pigment composition is controlled by the intensity (photon flux density) and quality

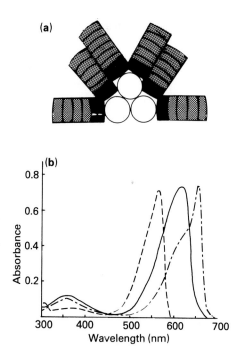

Fig. 1–4 (a) A three dimensional model of the phycobilosome showing the arrangement of allophycocyanin (unshaded), phycocyanin (black) and phycoerythrin (stippled) units (after Bryant *et al.*, 1979, *Arch. Microbiol.*, **123**, 113–27. (b) Absorption spectra of allophycocyanin (— · — · —), phycocyanin (———) and phycoerythrin (– – – –) from *Fremyella diplosiphon* (after Bennett and Bogorad, 1971, *Biochemistry*, **10**, 3625–35).

(spectral composition) of light, and it has been suggested that the resulting changes enable the most efficient utilization of available light energy. The production of all major pigments and even the formation of thylakoids may be affected. In blue-greens phycobiliprotein synthesis is particularly susceptible to environmental influences. Change in pigmentation is largely attributable to a change in the phycocyanin:phycoerythrin ratio. Certain strains will appear distinctly blue-ish green when grown in red light because phycocyanin, which absorbs mainly red light, is present in the greatest quantities; whilst if they are grown under a green light regime, they will appear rather reddish due to the predominance of the green light absorbing phycoerythrin in their cells. Thus they produce mainly the pigment which absorbs efficiently the light present. The shift between the red light and green light effects was found to occur at about 590 nm wavelength. This phenomenon has been called 'complementary chromatic adaptation'. It can be observed only in those blue-greens which synthesize phycoerythrin but not all phycoerythrin-producing blue-greens possess the ability to adapt chromatically.

Phycoerythrin synthesis occurs at maximum rate in green light and is reduced or completely suppressed in red light (Fig. 1–5). The action maximum for the induction of phycoerythrin synthesis in green light is at a wavelength of about 540 nm. Red light with an action maximum at about 650 nm reverses the induction of phycoerythrin synthesis. Rifamycin, an inhibitor of RNA polymerase, prevents subsequent phycoerythrin synthesis. This indicates that chromatic adaptation is genetically controlled at the level of DNA transcription. The control is mediated by photoreceptive pigments called 'adaptochromes', which appear to be closely associated with the phycobilisomes; they sample the quality of light and induce the changes in pigment synthesis. Adaptochromes are probably present in two forms, which interchange reversibly, one absorbing green light and the other red light; thus they display green versus red antagonism, in contrast to the red/far-red reversible transformations characteristic of higher plant phytochromes. Phycocyanin synthesis in some species is independent of

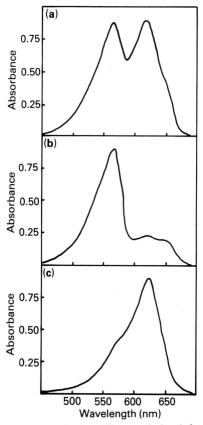

Fig. 1–5 Absorption spectra of phycobilisomes isolated from strain LPP–7409 following growth in white (**a**), green (**b**) or red (**c**) light (after Bryant *et al.*, 1979, *Arch. Microbiol.* **123**, 113–27).

the spectral composition of light, in others it may be under chromatic control, occurring at maximum rate in red light and at its lowest rate in green light. Allophycocyanin synthesis appears to be uneffected by light quality.

Three groups of blue-greens can be recognized according to their response to the spectral composition of light. In the first group phycoerythrin is always present and its synthesis is increased in green light; phycocyanin levels are unchanged. In the second group phycoerythrin synthesis is induced in green light and suppressed in red light; phycocyanin is always present and its synthesis is enhanced in red light. In strains belonging to the third group, neither phycoerythrin synthesis nor phycocyanin production are affected by the spectral composition of light.

1.6 Cytoplasmic inclusions and reserve products

Though light microscopy is inadequate for resolving the subcellular structure of blue-greens, it nevertheless shows clearly that their cytoplasm is heterogenous and usually incorporates granular structures of considerable density. Electron microscopic examination implies that these optical features arise from the presence of various cytoplasmic inclusions (Fig. 1–6). The most prominent structures are:

(*i*)   Glycogen granules, which are minute (about 30×65 nm in size), ovoid or rod-shaped structures deposited primarily in the cytoplasm between the thylakoids, and serve as a carbon and energy source.

(*ii*)   Lipid globules, which are spherical, variable in size (30 to 90 nm in diameter), most abundant near the cell surface, and possibly represent lipid stores for use in membrane synthesis.

(*iii*)   Cyanophycin (or 'structured') granules are relatively large in size (up to 500 nm in diameter), and are also deposited mainly at the cell periphery. They consist of a unique polypeptide (multi-L-arginyl poly-L-aspartic acid) composed only of two amino acids, arginine and aspartic acid. The arginyl residues are attached at each free carboxyl group of the poly-aspartate core. They are produced, unlike other polypeptides, by a ribosome-independent mechanism, and function in nitrogen turn-over.

(*iv*)   Polyphosphate bodies, which are fairly large structures and appear highly electron-dense even in thin sections, serve as a phosphate store.

(*v*)   Carboxysomes (or polyhedral bodies) are semicrystalline structures about 200–300 nm in diameter. They contain a reserve form of the primary photosynthetic enzyme, ribulose-1,5-bisphosphate carboxylase (RuBPCase) which, in its active form, catalyzes photosynthetic CO_2 fixation into ribulose-1,5-bisphosphate (RuBP).

Several other inclusions (like microtubules, microfilaments, wall bodies, crystalline and trilamellar bodies, and also granular and fibrillar crystals) have been observed in thin sections from various blue-greens but so far nothing is known about their function.

It may well be that the success and competitive powers of blue-greens in natural habitats are, at least partly, due to their remarkable ability to store essential nutrients and metabolites within their cytoplasm. Reserve products are accumulated under conditions of excess supply of particular nutrients (orthophosphate in case of polyphosphate granules or combined nitrogen for cyanophycin granules). Conversely, accumulation of storage products may be induced under conditions of particular nutrient deficiency. For example, when the synthesis of nitrogenous cell constituents is halted in the absence of utilizable nitrogen source, the primary products of photosythesis are channelled towards the synthesis and accumulation of glycogen and lipid. Similarly large reserves of polyphosphate are laid down when phosphate supplies are restored to phosphorus-starved cells.

Reserve products are mobilized during periods of nutrient shortage or under conditions when photosynthesis is not possible. Polyphosphate bodies are depleted in the absence of phosphate in the environment, cyanophycin granules are degraded under nitrogen-limited growth conditions, and glycogen granules are utilized as a source of carbon in biosynthetic processes or for energy prodution in the dark. Glycogen accumulated during nitrogen depletion is degraded upon fresh supply of a nitrogen source thus providing carbon skeletons to incorporate the nitrogen assimilated.

The accumulation and degradation of reserve material is catalyzed by specific enzymes. Cyanophycin polypeptide is largely absent in exponentially growing cell material, it accumulates in stationary phase cultures, and is rapidly degraded during re-newed growth when cells are transferred into fresh nutrient medium. When protein synthesis is inhibited at the transcriptional level, the nitrogen assimilated will be channelled towards the production of cyanophycin polypeptide even in a logarithmic phase culture. The enzyme catalyzing cyanophycin synthesis appears to be always present in the cells and is independent of transcriptional control. The cyanophycin-degrading enzyme is, however, not constitutively present; its synthesis is induced when required.

Phycobiliproteins are present often at very high concentrations in the cells of blue-greens, constituting up to 40–60% of the cell protein. In addition to their primary function in photosynthesis, they may also serve as reserve protein. Under conditions of nitrogen limitation cellular phycobiliprotein content may decrease by half or more, being utilized as a nitrogen source for the synthesis of other essential proteins and nitrogenous cell consitituents. Degradation of phycobiliproteins is controlled by specific proteases which have been demonstrated in extracts from blue-greens.

When nitrogen-depleted cells are provided with a suitable source of nitrogen, the nitrogen assimilated is first utilized for the rapid synthesis of large quantities of cyanophycin polypeptide. Subsequently there is an equally speedy decrease in the content of cyanophycin while enzymes and phycobiliproteins are being synthesized. Similarly, large amounts of glycogen are formed during a period of nitrogen starvation, and are mobilized upon the supply of utilizable nitrogen. The ability of rapid uptake of essential nutrients and their swift accumulation in the form of greatly concentrated reserve material, and the eventual controlled mobilization of reserves, is an extremely important adaptation of blue-greens to their rapidly changing physical environment.

1.7 Gas vacuoles

The strongly refractive appearance of the cells of planktonic blue-greens under the light microscope is due to the presence of gas-filled cylindrical vesicles. Many hundreds of vesicles are assembled in tight groups or bundles in the cytoplasm seen as bright areas of the cell called gas vacuoles (Fig. 1–6b). A

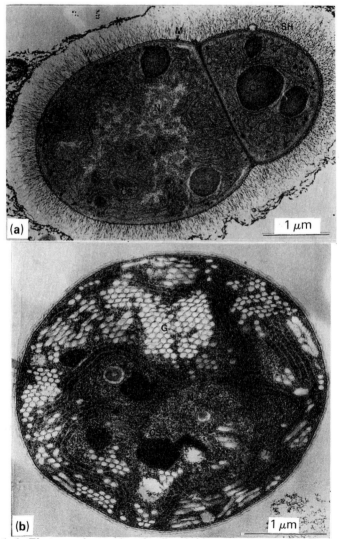

Fig. 1–6 Electron micrographs of cells of (**a**) *Anabaena* and (**b**) *Microcystis*. SH, sheath; W, cell wall; M, cell membrane; N, nucleoplasm; T, thylakoid; CP, cyanophycin; CS, carboxysome; G, gas vesicle. By courtesy of Dr L.V. Leak (a) and Dr M. Jost (b).

single gas vesicle is a cylinder (70 nm in diameter and about 350 nm long) closed at both ends by conical caps. The vesicle is formed by an extremely thin (2 nm) and rigid proteinaceous membrane enclosing a gas-filled space. The membrane is built of protein molecules that are tightly packed in rows along the ribs that form the vesicle. The membrane is porous and freely permeable to gas molecules. Consequently the content of gas vesicles is determined by the nature of gases dissolved in the surrounding cytoplasm and in the external medium.

Gas vesicles are delicate structures and tend to collapse when the cell is subjected to a pressure of several atmospheres. Their collapse was demonstrated in 1895 by Klebahn in his famous 'hammer, cork and bottle' experiment. He filled a bottle with a suspension of buoyant gas-vacuolate blue-greens and corked it. Striking the cork with a hammer caused a dramatic change of the suspension from a milky green to a dark translucent green. Subsequently gas bubbles collected under the cork and the cell material sank to the bottom of the bottle (Fig. 1–7 a and b). Phase-contrast light microscopic examination has shown that the pressurized cells lose their refractive property (Fig. 1–7 c and d). The actual collapse of gas vesicles has been established by electron microscopy.

Gas vesicles are the buoyancy-regulating organelles and responsible for the diurnal vertical migration of certain planktonic blue-greens. This movement enables them to maintain their position within a zone of the water column where their photosynthetic metabolism is optimal and the supply of mineral nutrients adequate. The mechanism of buoyancy regulation and its ecological significance are discussed in more detail in section 4.1.

1.8 Surface structures

The protoplast of blue-greens is surrounded by a cell wall, the construction and chemical composition of which is similar to the wall of Gram-negative bacteria. The wall is composed of two principal layers, the inner murein or petidoglucan and the outer lipoprotein layers (Fig. 1–6 a and b). The murein layer which forms a polymeric fibrous mesh is primarily responsible for the mechanical strength of the wall; it confers the shape of the cell and protects it against osmotic damage. The lipoprotein layer probably controls the transport of solutes as in bacteria. The space between the two layers, termed the perplasmic space, appears to have a similar content of lipopolysaccharides and degradative enzymes as in Gram-negative bacteria. The presence in the periplasmic space of alkaline phosphatase (effecting the release of phosphate residues in organic combinations) has been demonstrated in several strains of blue-greens. Electron micrographs of thin sections of various filamentous blue-greens have revealed the presence of fine (about 20 nm in diameter) channel-like structures, termed microplasmodesmata, that traverse the cross walls between adjacent cells; they extend to the plasma membranes of the two cells and probably function in establishing intercellular contact.

The cell wall is often covered by a voluminous fluffy or dense, sometimes laminated sheath or capsule (also termed 'glycocalyx'). It is essentially a slime

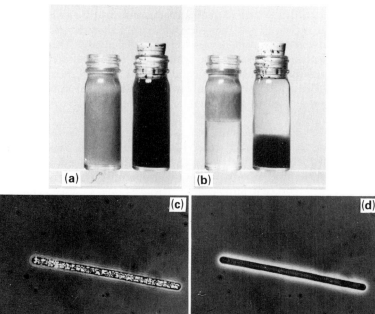

Fig. 1–7 The hammer, cork and bottle experiment showing the decrease of turbidity after striking the cork (**a**), and the loss of buoyancy after allowing the bottle to stand for 2 hours (**b**). A filament of *Oscillatoria* with refractive gas vacuoles (**c**), and after the collapse of gas vesicles (**d**). By courtesy of Prof. A.E. Walsby.

layer composed of polysaccharide and incorporates fine fibrillar elements (see Fig. 1–6a). The function of this envelope is poorly understood but it has been suggested that it may protect the cells from desiccation, or against invasion of pathogens or against phagocytosis by grazing animals. It may also render toxic or harmful substances ineffective and promote the attachment of the organisms to solid substrates. Large quantities of metabolic products may be invested in the formation of extensive sheaths. Hence it is not surprising that the synthesis of sheath polysaccharides is under genetic control. The ability to produce sheaths may be lost upon repeated transfers of blue-greens in culture. It is possible that the factors in nature which induce sheath formation are no longer present in culture and hence the synthesis of sheath material becomes wasteful and may be disadvantagous. A spontaneously produced sheathless mutant strain will therefore outgrow the wild sheathed strain in culture.

Parallel arrays of fine fibrils, about 7 nm in diameter and 1–2 μm long, have been detected at the cell surface of several blue-greens by electron microscopy. It has been suggested that these slender fibrils may be involved in gliding movement (for details see § 4.7). Similar structures of as yet unknown function were observed on the cell surface of endosymbiotic blue-greens strains, and have been compared to the fimbriae (or pili) of Gram-negative bacteria.

2 Patterns of Development and Reproduction

We may conclude from the previous chapter that blue-greens exhibit a greater degree of morphological and structural complexity than any other group of Prokaryota. Correspondingly, they also show more diversity in their modes of reproduction and in their patterns of development. Furthermore, cellular differentiation for a specific metabolic function, e.g. the formation of nitrogen-fixing cells (heterocysts), is without parallel among the prokaryotic micro-organisms.

2.1 Heterocyst differentiation

In the process of nitrogen fixation atmospheric nitrogen (N_2) is reduced to ammonia (NH_3) in a reaction catalyzed by nitrogenase, a complex enzyme system. An absolute condition for this reaction is the absence of free O_2 since nitrogenase is inactivated in the presence of O_2. It is thought that the ability to fix N_2 developed during an early period of biological evolution when reducing conditions were manifest in the environment. When later the atmosphere became increasingly oxygenic, there must have been a need to protect nitrogenase by the removal or exclusion of free O_2 from the site of enzyme action. One of the most efficient devices acquired during the evolutionary process was the heterocyst.

Heterocysts are produced under nitrogen-depleted conditions by the gradual transformation of certain vegetative cells. Their formation is suppressed in the presence of combined nitrogen, particularly ammonium–nitrogen, in the environment. Under these conditions undifferentiated trichomes are produced with a uniform cell population. Heterocyst differentiation is induced by the exhaustion or deprivation of combined nitrogen in the medium. The transformation involves a whole range of structural and biochemical changes, including the mobilization of reserve products, the re-organization of the cytoplasmic membrane system, protein degradation and the synthesis of new proteins, the deposition of a multilayered envelope, and the formation of a tubular neck, the pore channel, between heterocyst and adjacent vegetative cell (Fig. 2–1). An amorphous body (the 'plug') is deposited in the pore channel of mature heterocysts which contains arginine and consists almost certainly of cyanophycin polypeptide. Its possible function in nitrogen metabolism is discussed in section 3.7.

During the initial period of nitrogen limitation, the ratio of carbon to nitrogen (C:N) of the trichome rapidly increases from a value of about 4.5 to 8.0, as nitrogenous reserves (cyanophycin, phycobiliproteins) become depleted (Fig. 2–2). Thus a severe state of nitrogen starvation, which is

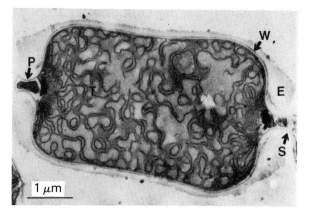

Fig. 2–1 Electron micrograph of a heterocyst of *Anabaena*. E, envelope; W, cell wall; S, septum; P, pore channel; T, thylakoid. By courtesy of Prof. N.J. Lang.

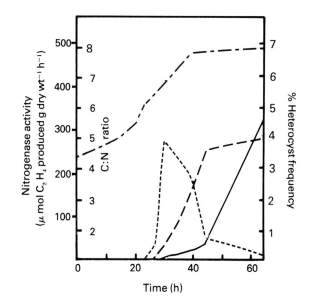

Fig. 2–2 Course of heterocyst differentiation (- - - - - - - proheterocysts; — — — —, heterocysts) and of nitrogenase activity (————) in relation to changes of cellular C:N ratio (— - — - —) of *Anabaena cylindrica* (after Kulasooriya *et al.*, 1972, *Proc. R. Soc. B*, **181**, 199–209).

reflected in the high cellular C:N ratio, seems to be the internal stimulus which in some way triggers off the process of heterocyst differentiation and the accompanying synthesis of nitrogenase. The transforming heterocyst shows high proteolytic activity, and as a result of protease action about half of the proteins originally present in the undifferentiated cell are degraded during heterocyst development. Subsequently new proteins, such as the nitrogenase enzymes, are synthesized in the final stage of cell differentiation. The peripheral photosynthetic thylakoids disintegrate and are replaced by a new elaborate and confluent membrane system which extends throughout the cytoplasm of the heterocyst. A massive multilayered envelope is deposited on the external surface of the original cell wall, which enshrines almost the whole heterocyst, apart from the much reduced areas of contact between the heterocyst and vegetative cell(s). The much reduced septum is traversed by fewer fine channels (microplasmodesmata) than the cross wall between two vegetative cells (approximately 50 against 250); it is assumed that they represent the routes of metabolic interchange between the heterocyst and its neighbouring cells. The external ('fibrous' and 'homogenous') envelope layers are composed of polysaccharide but the innermost 'laminated' layer contains complex glycolipids which are thought to act as permeability barriers to the movement of molecules into the heterocyst, thereby controlling, but not preventing, the diffusion of gases. It is, however, possible that controlled gaseous exchange takes place solely through the microplasmodesmata.

Heterocyst retain significant quantities of chlorophyll a and carotenoid pigments but essentially lack phycobiliproteins. While Photosystem I units are increased, much of the chlorophyll and phycobiliproteins associated with Photosystem II units in vegetative cells are lost during differentiation (see § 3.1). Heterocysts are also devoid of the enzyme ribulose-1,5-bisphosphate carboxylase and of manganese (Mn^{2+}), an essential component of Photosystem II. The consequence of these changes is a modified photosynthetic mechanism which no longer performs photosynthetic CO_2 fixation or O_2 evolution. On the other hand, heterocysts possess all the principal Photosystem I components, like chlorophyll P700, cytochrome–554 and 563, ferredoxin and ferredoxin–NADP (nicotinamide adenine dinucleotide phosphate)–oxidoreductase, plastoquinone and plastocyanin. They have been shown to be capable of photophosphorylation (the synthesis of ATP, adenosine triphosphate, in the light) and of NADP photoreduction. These Photosystem I activities provide the required energy for N_2-fixation in heterocysts (see Fig. 2–3 and § 3.7).

Heterocysts maintain strong reducing conditions; they rapidly reduce redox indicators such as tetrazolium salts. Isolated heterocysts show high rates of endogenous respiration. Activities of enzymes involved in dark energy metabolism are several times greater in heterocysts than in vegetative cells. Respiratory activities are probably associated with the plasmalemma and the elaborate membrane system of heterocysts, and may function both in the protection of nitrogenase, through removal of O_2 (see § 3.7), and in generating ATP and reductant by means of respiratory electron transport.

It is now clear that all the structural and biochemical changes which take place during heterocyst differentiation serve to promote the prime function of heterocysts in N_2-fixation. This function was for long obscure, and it was only in

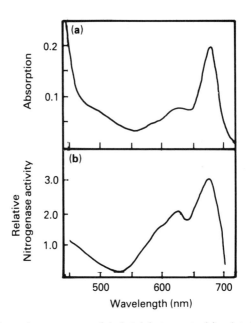

Fig. 2–3 Absorption spectrum of isolated heterocysts (**a**) related to the action spectrum of nitrogenase activity (acetylene reduction) by intact filaments (**b**) of *Anabaena cylindrica* (after Fay, 1970, *Biochim. biophys. Acta*, **216**, 353–6).

1968 that sufficient information had gathered to propose that heterocysts are the site of N_2-fixation. This was shortly followed by the demonstration of N_2-fixation activity in isolated heterocysts.

2.2 Control and patterns of heterocyst development

Heterocysts are formed at regular intervals along the trichomes of nostocacean blue-greens. As the trichome elongates through cell growth and division, and the distance between two heterocysts increases, a new heterocyst ('proheterocyst') begins to differentiate about midway between the existing two heterocysts. Considering an average heterocyst frequency of about 5%, it would mean that about every twentieth cell will become a heterocyst. Such an even distribution of heterocysts is evidently desirable. It is also reasonable to assume that in general a single heterocyst is capable of providing about twenty vegetative cells with fixed nitrogen. The regular spacing of heterocysts is obviously an efficient way of distributing the fixed nitrogen in a linear multicellular organism. The heterocyst-vegetative cell interrelationship involves the exchange and two-way transport of essential metabolites along the trichome. The movements of carbon into heterocysts and of nitrogen out of them have been demonstrated by means of pulse radioactive labelling followed by autoradiography. Vegetative cells provide the heterocysts with organic substances required partly as a source of energy and reductant in N_2-fixation, and partly to bring the freshly fixed nitrogen (NH_3) into organic combination.

The question of how such a regular pattern is reproduced has long puzzled researchers. Fogg, who in the 1940s undertook the first systematic studies on N_2-fixation and heterocyst production in *Anabaena cylindrica*, proposed that heterocyst formation is induced when the concentration in a vegetative cell of a specific nitrogenous inhibitory substance, probably NH_3 or a simple derivative of it, falls below a critical level. He also provided evidence that the source of the inhibitory substance is the heterocyst itself, and he postulated that this substance is utilized in vegetative cell metabolism, while it diffuses from cell to cell along the trichome, producing a decreasing concentration gradient away from the heterocyst. When undifferentiated filaments (grown in the presence of NH_4^+-nitrogen) are deprived of combined nitrogen, and the process of heterocyst differentiation begins, proheterocysts arise at regular interval thereby establishing the characteristic heterocystous pattern before heterocysts are fully developed and begin to fix N_2. Indeed, the pattern is reproduced even when N_2-fixation is prevented by the exclusion of atmospheric N_2 (e.g. by incubating the filament suspension under an artificial gas phase composed of argon and CO_2). It is therefore possible that some products of protein degradation, in addition to the products of N_2-fixation, are the source of the diffusible inhibitor, and as this is released from the proheterocysts, it sets up a zone of inhibition around neighbouring vegetative cells. The concentration of the inhibitor will fall to a critically low level at a certain distance from the heterocyst, permitting the formation of a new heterocyst. The identity of the inhibitor, which also may be involved in the repression of nitrogenase synthesis, is uncertain. Recent studies have shown that other factors, in addition to the diffusible inhibitor, may play in establishing the heterocystous pattern. One concerns the history of the cell giving rise to the heterocyst. In *Anabaena catenula* vegetative cell division is unequal or asymmetric and results in the production of a smaller and a larger daughter cell. Heterocysts always develop from the smaller daughter cell of a division which occurred outside the inhibitory zone of an existing heterocyst. It was also shown that a cell is only susceptible to the stimulus to differentiate into a heterocyst during an early stage of its developmental cycle. All these prerequisites of heterocyst formation place constraint upon the population of cells within a trichome, thus limiting the number of cells which may transform into a heterocyst, and establishing thereby the characteristic pattern of a heterocystous filament.

The heterocystous pattern may show some variation in other groups of heterocyst-forming blue-greens (Rivulariaceae, Stigonematales). In *Calothrix*, for example, the initial response to nitrogen depletion is similarly the formation of intercalary heterocysts at more or less regular intervals. This is subsequently followed by the breakage of the trichome near the heterocyst. The end result is a polar trichome with a single terminal heterocyst. Although the final pattern differs, the primary factors regulating heterocyst development are probably similar in all groups of heterocystous blue-greens.

2.3 Cell division

The simplest mode of propagation is by cell division which is followed by growth and expansion. Cell division (binary fission) may be the sole mode of reproduction in the unicellular blue-greens. It constitutes the basic mode of growth in the multicellular forms.

Cell division in a single plane perpendicular to the long axis of the ovoid or rod-shaped cell (transverse binary fission) is characteristic for the genera *Synechococcus* (*Anacystis*, *Agmenellum*, *Aphanothece*, *Coccochloris*) and *Gloeothece*). Unicellular forms distinguished by their spherical shape and assigned to the genera *Synechocystis* (*Aphanocapsa*, *Microcystis*) and *Gloeothece* (*Gloeocapsa*). Unicellular forms distinguished by their spherical shape and assigned to the genera *Synechocystis* (*Aphanocapsa*, *Microcystis*) and perpendicular plans, and the cells as a rule elongate in a direction perpendicular to the long axis of the ovoid mother cell. Such synchronous cell divisions give rise to an ordered rectangular plate-like colony; the cells are being held together by a secreted membraneous sheath (Fig. 1–1c). The cubical colonies of *Eucapsa* arise through cells dividing successively in three planes at right angles to each other, and the daughter cells expanding in a direction parallel to the previous plane of cell division.

Cell growth and division are closely related to nucleic acid synthesis. DNA replication is completed prior to cell division. The rate of RNA synthesis is, however, highest during cell growth between two cell divisions. Under normal conditions cell growth and cell division alternate regularly.

Growth and cell division are closely linked also in filamentous forms, and the process of cell division is basically similar to that described in unicellular blue-greens. The main difference is the lack of separation of the daughter cells. A more complex situation has developed in certain filamentous forms in which growth and cell division are not completed in a single cycle but occur almost simultaneously and continuously. New cross walls are initiated before developing septa are fully completed during the seemingly steady state of DNA replication and cell division.

2.4 Modes of reproduction

Most unicellular blue-greens reproduce by binary fission. An exception is seen in the genus *Chamaesiphon* where a new cell is produced in the process of an unequal fission or budding (Fig. 1–1e). Spherical cells are budded off at one end (the apical pole) of the ovoid or elongated sessile vegetative cell, which displays a distinct, basal to apical, polarity.

In the order Pleurocapsales a characteristic mode of reproduction is by multiple fission (Fig. 1–1f). This occurs through a series of rapid divisions in three planes of the enlarged mother cell, and is not acompanied by cell growth. The small spherical baeocytes produced are released by the rupture of the mother cell wall.

There are various modes of reproduction in the filamentous forms, and some involve the formation and germination of specific reproductive structures. The simplest and in some genera the commonest way of reproduction is by random trichome fragmentation. The trichome fragments may vary in size but even a single detached cell may be capable under favourable conditions of reproducing the characteristic filamentous morphology of a particular species. Evidently the relevant genetic information is present in every cell. Filament fragmentation is often a more ordered process associated with the production of distinct reproductive trichome segments called hormogones (or hormogonia). These are short 5 to 15 celled segments which separate by the rounding off of their end cells within the trichome envelope. They exhibit active gliding motion (see § 4.7) upon their liberation, and develop into a new trichome. Hormogone development in certain species is initiated by the formation along the trichome of separation discs ('necridia'). These are cells which undergo lysis and dehydration, and serve as preformed breaking points for hormogone detachment. While in some genera, like *Oscillatoria*, several hormogonia may be produced along a single trichome, in others hormogonia develop in more defined positions, for example below the apical 'hair' in *Rivularia* or at the end of the lateral branches in *Fischerella*.

2.5 Akinetes

The various modes of reproduction discussed in the previous section are typical of actively-growing populations and serve the dissemination of species under circumstances favourable for growth. Many of the heterocystous blue-greens are also capable of producing perennating structures, akinetes or spores, under adverse environmental conditions which develop through the transformation of vegetative cells (Fig. 2–4a).

During the process of akinete differentiation the cell increases several fold in size, accumulates large reserves of cyanophycin polypeptide and glycogen, and deposits a thick and elaborate extracellular capsule which completely surrounds the mature akinete (Fig. 2–4b). The morphological-structural changes are accompanied by profound changes in the biochemistry and physiology of the cell. Polyphosphate disappears, pigment composition changes, and RNA content is greatly increased. In addition, photosynthetic and biosynthetic activities are reduced and may cease altogether upon completion of akinete differentiation but respiratory activity remains relatively high. Though the spatial distribution of akinetes within the trichome shows a variety of patterns, in all cases akinetes develop consistently in a particular position with respect to the heterocyst. Akinetes always form adjacent to the terminal heterocyst in the genus *Cylindrospermum*, they develop from the vegetative cell 2–3 cells away from a heterocyst in most planktonic *Anabaena* species, and are initiated in cells approximately equidistant between two heterocysts in *A. doliolum* and in many *Nostoc* species. In the latter akinetes are often produced synchronously in chains. In some heterocystous genera (*Calothrix, Mastigocladus*) akinetes are never observed.

The consistent spatial relationship between heterocysts and akinetes seems

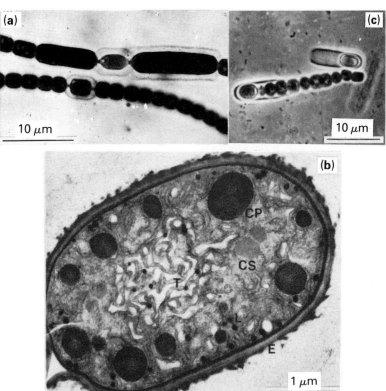

Fig. 2–4 (a) Heterocyst flanked by akinetes of *Anabaena* (U.V. micrograph).
(b) Akinete of *Anabaenopis* (electron micrograph). T, thylakoid; CP, cyanophycin;
CS, carboxysome; L, lipids; E, envelope. (c) Germinating akinete of *Anabaena* (light
micrograph). By courtesy of Prof. G.E. Fogg (a), and Prof. N.J. Lang (b).

to suggest that heterocysts may exert some degree of control over akinete
formation, though it appears that in addition to the influence of heterocysts,
some other external factors play a part in triggering the process of sporulation.
In temperate lakes and reservoirs, akinete production coincides with the
accumulation of planktonic populations during the formation of dense water
blooms (see § 4.1.2), under conditions which are unfavourable for vegetative
growth and which resemble those prevalent in post-exponential phase cultures.
Many suggestion have been put forward to explain how akinete formation is
induced but none of the explanations appears to be generally applicable.
Recently increased population density leading to decreased light penetration
and energy limitation has been thought to be a more important condition in
initiating akinete development than nutrient deficiency. The nature and the
mode of action of the immediate biochemical trigger which controls the onset
of akinete differentiation (which is responsible for the cessation of cell division,
for the excessive growth and accumulation of reserve products, and deposition

of a massive envelope) is as yet unknown. Neither do we have adequate factual information regarding the part heterocysts may play in the regulation of akinete formation to explain the distinct spatial relationship between akinetes and heterocysts and the reasons behind the development rest of various akinete patterns.

In laboratory cultures akinete formation is often closely followed by akinete germination indicating that no obligatory lag phase (or resting period) is necessary before the onset of germination. Only a small proportion (10 to 15%) of the akinetes formed, however, will germinate under the increasingly unfavourable conditions that develop in post-exponential cultures. The majority of akinetes will become detached from the filament and settle to the bottom of the culture vessel. Most of them will, however, readily germinate upon transfer into fresh nutrient medium.

The process of akinete germination begins with gradual disappearance of reserve products, and this is accompanied by the synthesis of new DNA, RNA, pigment and protein and the formation of new thylakoids. These are followed by one or two transverse divisions, and finally by the local dissolution of the akinete envelope, enabling the emergence and release of the short germling (Fig. 2–4c). In some strains cell division will not begin before the emergence of the germinating cell. The germling displays a jerking or a rhythmic gliding movement on a solid substratum. In the absence of combined nitrogen one of the terminal cells usually develops into a heterocyst.

Akinete formation and germination may have an important role in the survival of planktonic blue-greens which frequently undergo photo-oxidation and rapid breakdown in surface blooms (see § 4.1.2). Akinetes were reported to survive several years of deposition in the anaerobic mud sediments and to be capable of germination after many years of storage in herbarium specimens.

2.6 Life cycle and photoregulation of development

Observations of the growth of blue-greens in nature and in culture suggest that reproductive structures (like baeocytes, hormogonia or akinetes) are produced at some stage in their growth. Distinct alternation of morphological forms may occur during the developmental cycle of most colonial and filamentous genera, though detailed investigation of developmental sequences have been undertaken with relatively few species.

In the plaeurocapsalean genus *Dermocarpa* the emerging minute baeocyte gradually develops into a vegetative cell. Its maturation is accompanied by about a ten-fold increase in size. The fully developed cell then gives rise by means of cytoplasmic cleavage to a large number of new baeocytes (Fig. 1–1f). Multiple fission is preceded by multiple nuclear division. The liberated baeocytes display gliding motility, and respond by way of phototactic movement to a light gradient. They eventually come to rest after a motile phase usually lasting a few hours before developing into a vegetative cell.

In the genus *Dermocarpella* the developmental process is characterized by a distinct polarity. Here the baeocyte enlarges asymmetrically to develop into a pear-shaped vegetative cell, which when fully mature divides unequally into a small basal vegetative cell and a large apical reproductive cell. The vegetative cell remains closely attached to the mother cell envelope. The reproductive cell produces a large number of baeocytes by means of rapid successive cell cleavage. They are set free by the rupture at the apical end of the parental cell envelope. Subsequently the vegetative cell enlarges and its asymmetric cell division signals the onset of a new developmental cycle.

Developmental sequences in filamentous forms and particularly in some heterocystous genera are more complex. During the well documented development of *Nostoc muscorum* two separate cycles, the heterocystous and the sporogenous cycles can be distinguished (Fig. 2–5). When motile

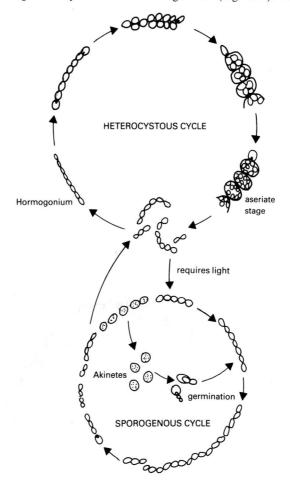

Fig. 2–5 Alternation of heterocystous and sporogenous generations in *Nostoc muscorum* (after Lazaroff and Vishniac, 1962, *J. gen. Microbiol.*, **28**, 203–10).

hormogonia produced in the heterocystous cycle come to rest, the terminal cell of the juvenile filament develops into a heterocyst, whilst the remaining cells continue to grow and divide, each giving rise to cluster of four cells surrounded by the parent envelope: this is referred to as the 'aseriate' (cells not in a line) stage of development. When the organism is grown on organic substrates in the dark (see § 3.4), the process is arrested at this stage. Further development requires illumination, but it may continue in the dark after only a brief exposure to light. Light activation induces the liberation of the cell clusters from the mother cell envelope, which emerge in the form of short motile trichomes. They grow into hormogonia and later form intercalary heterocysts. Their release is the beginning of a new heterocystous cycle. The developmental process in *N. muscorum* may take an alternative sequence when the heterocystous trichomes undergo excessive sporulation. This is called the sporogenous cycle. The akinetes eventually germinate and the emerging germlings continue their development through the heterocystous cycle.

Photo-induction of filamentous development in the Nostocaceae depends on the quality of light. Red light of about 650 nm wavelength promotes the formation of motile trichomes, whilst the process of induction is reversed by exposure of trichomes to green light. The photoreversibility of this process indicates the involvment of a phytochrome type photomorphogenetic pigment or pigment system. It is now well established that photomorphogenetic responses as well as chromatic adaptation are mediated by photoreversibly photochromic pigments (adaptochromes; see § 1.5), most probably related to phycobiliproteins and located in phycobilisomes. The actual photoreceptor has not yet been isolated or characterized; it could be identical with any of the three main phycobiliproteins or could be the photochromic modification of one of them.

3 Metabolism

Some of the more conspicuous features of blue-greens described in the previous chapters, like their coloured appearance, the high concentration of light-absorbing pigments in their cells and the highly developed intracytoplasmic membrane system, are unmistakable indications that photosynthesis is the principal mode of energy metabolism in these organisms. In the natural environment, however, blue-greens regularly experience dark night periods, and it is known that some are able to survive long periods in complete darkness. This suggests that they possess a respiratory metabolism capable of providing maintenance energy in the dark. Furthermore certain strains show a distinct ability for heterotrophic nutrition.

3.1 Photosynthesis

Photosynthesis can be defined as the synthesis of organic compounds through the assimilation ('fixation') of CO_2 with the use of light as an energy source. CO_2 is incorporated into a 5-carbon acceptor, ribulose-1,5-bisphosphate (RuBP) in an energy requiring reaction catalyzed by the primary carboxylating enzyme, RuBP carboxylase (Fig. 3–1). The product splits into two molecules of a 3-carbon compound, phosphoglyceric acid (PGA), and the reduction of PGA, mediated by the electron carrier NADPH (nicotinamide adenine dinucleotide phosphate) leads to the formation of a series of sugar phosphate intermediates and finally to glucose. During this sequence of metabolic transformations, known as the reductive pentose ·phosphate pathway (or Calvin cycle), the acceptor RuBP is regenerated, ready to accept another CO_2 molecule.

Blue-greens, unlike green algae, are able to utilize both free CO_2 and bicarbonate ions as a source of inorganic carbon in photosynthesis. Planktonic blue-greens therefore possess a distinct competitive advantage over other planktonic forms in natural water of high pH and high alkalinity. Bicarbonate ions are transported in the light across the plasma membrane and accumulate in the cell to serve as an inorganic carbon pool for photosynthesis. Bicarbonate is converted to CO_2 by the action of the enzyme carbonic anhydrase:

$$HCO_3^- + H^+ \rightarrow CO_2 + H_2O$$

Carbonic anhydrase activity may rapidly increase when external CO_2 concentrations suddenly decrease.

Fixation of CO_2 is not directly dependent on light, and the process is therefore termed the 'dark reaction' of photosynthesis. The requirements for energy (in the form of ATP) and reductant (NADPH), however, render the

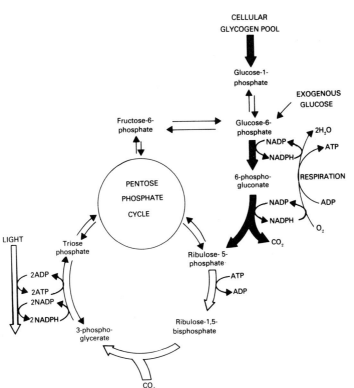

Fig. 3–1 Pathways of light and dark carbon metabolism in blue-greens. Reactions specific to photoautotrophic metabolism are indicated by white arrows, those specific to dark respiratory metabolism by black arrows (after Pelroy *et al.*, 1972, *Arch. Microbiol.*, **87**, 303–22).

transformations during the Calvin cycle fully dependent on the primary photochemical act which takes place in the thylakoid membranes. It is here where light energy, absorbed by the highly organized assemblies of photosynthetic pigments and electron carriers, called Photosystem I and Photosystem II, excites the chlorophyll molecule in the 'reaction centres'. This leads to the expulsion of high energy electrons and their flow down a redox potential gradient which results in the formation of strongly electro-negative electron carriers, like ferredoxin and NADPH (Fig. 3–2). Part of the released energy is incorporated during this electron transport into ATP in the process of photophosphorylation. The ultimate source of electrons for photosynthesis is water, which yields in the process of photolysis (Hill reaction) hydrogen atoms (protons), electrons and free O_2, an important by-product of blue-green and all green plant photosynthesis:

$$H_2O \rightarrow 2H^+ + \frac{1}{2}O_2 + 2e^-$$

Thus the main characteristics of photosynthesis, originally elucidated in green algae and higher plant chloroplasts, also apply to blue-greens. There are two important aspects of photosynthesis which are, however, particular to blue-greens and warrant attention here.

One concerns the spectral characteristics of light absorption in blue-greens which are inherently different from those of other photosynthetic organisms. This was noted as long ago as 1883 by Engelman, and investigated in detail some 60 years later. High rates of photosynthetic activity are measured in blue-greens not only in the spectral region between 665–680 nm wavelengths, where light absorption by chlorophyll is greatest, but also around 620 nm or 560 nm, where phycocyanin and phycoerythrin, respectively, absorb light effectively (see Figs. 1–3 and 1–5). It was shown that light absorbed by the phycobiliproteins is used by blue-greens as efficiently as light absorbed by chlorophyll. Excitation energy is transmitted from biliprotein pigments to chlorophyll with great efficiency. Energy transfer proceeds as a rule from a pigment with an absorption maximum at a shorter wavelength of light to a pigment with greatest

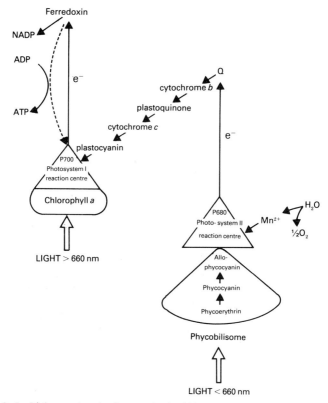

Fig. 3–2   Light reactions in photosynthesis of blue-greens.

light absorption at a longer wavelength of light. Hence the sequence is:
phycoerythrin → phycocyanin → allophycocyanin → chlorophyll *a*

Efficient energy transfer enables blue-greens to prosper under particular light regimes, like those encountered by aquatic forms deep below the water surface. Chromatic adaptation will further increase their photosynthetic efficiency. Although phycobiliproteins are not absolutely essential for either of the light reactions performed by Photosystem I or II, phycobiliprotein-deficient cells are much less competent in photosynthesis.

Another curiosity is the more recent finding that a group of non-heterocystous blue-greens is able to utilize ultimate electron donors other than water, like hydrogen sulphide (H_2S) or elemental hydrogen (H_2), for the reduction of CO_2 in a Photosystem I driven reaction and when Photosystem II was not operational (in the presence of the specific photosynthetic inhibitor dichlorophenyl dimenthylurea, DCMU). This anoxygenic type of photo-synthesis (not accompanied by O_2 evolution) is similar to that which operates in photosynthetic bacteria. H_2S is oxidized in the process to elemental sulphur which is deposited outside the cell:

$$2H_2S + CO_2 \rightarrow CH_2O + 2S + H_2O$$

The transition from oxygenic to anoxygenic photosynthesis requires a brief period of adaptation. Though the ability to perform facultative anoxygenic photosynthesis was demonstrated in *Oscillartoria limnetica* isolated from a hypersaline sulphide-rich lake and in several other strains, the majority of blue-greens tested are unable to utilize sulphide as a source of photoreductant. Moreover, sulphide can be extremely toxic, particularly under acidic conditions, and inhibits normal oxygenic photosynthesis.

The concentration of the carboxylating enzyme, RuBP carboxylase is in general exceptionally high in blue-greens, amounting up to 30% of the total cell protein. The enzyme is present both in soluble and particulate forms, and the ratio between the two fractions changes during growth. Maximum CO_2-fixation is observed when the soluble:particulate enzyme ratio is highest. This finding supports the assumption that the soluble fraction is the active form of the enzyme while the particulate fraction, which is present in the carboxysomes (see § 1.6) is not directly involved in the reduction of CO_2, and appears to represent a storage form of the enzyme.

Though carboxylation of RuBP is the main route of CO_2 incorporation in blue-greens under optimum photosynthesizing conditions, it is not the only pathway for CO_2-fixation. Carboxylation of phosphoenol pyruvate, catalyzed by the enzyme phosphoenol pyruvate carboxylase, represents another route for CO_2 uptake (Fig. 3–3). It has been shown to take place in various blue-greens in red light or in dim white light:

$$\text{phosphoenol pyruvate} + CO_2 \rightarrow \text{oxaloacetate} + PO_4^{3-}$$

Oxaloacetate is readily converted to C_4-dicarboxylic acids, e.g. to malate or citrate, and subsequently to amino acids, like aspartate or glutamate. This pathway, reminiscent of the C_4-dicarboxylic acid pathway (Hatch–Slack pathway) in higher plants, complements the reductive pentose phosphate pathway in blue-greens. The possession of two carboxylating systems, operating side by side, may represent an important adaptation of blue-greens to rapidly changing environmental circumstances. Under limiting light conditions carbon assimilation is preferentially channelled towards the synthesis of amino acids and other essential cell constituents, but under saturating light quantities, sugars and starch are formed via the reductive pentose phosphate pathway. This indicates that at high illumination the rate of carbon fixation may exceed the rate of nitrogen assimilation, and hence the excess carbon and energy derived from photosynthesis is stored in the form of glycogen.

3.2 Respiratory metabolism

Most blue-greens 'rest' in the dark. Dark endogenous metabolism serves mainly the adjustment of photosynthetic and biosynthetic mechanisms for the subsequent active light period. Glycogen is the main reserve product which supports a limited dark metabolism and provides maintenance energy required for the essential cellular processes in the dark. It is being converted to glucose-6-phosphate and metabolized via the respiratory pathways.

Although enzymes of the glycolytic pathway were detected in various blue-greens, their activities are extremely low, and it is therefore doubtful that fermentative metabolism could support anaerobic growth in the dark. Dark energy-yielding metabolism of blue-greens is distinctly O_2-dependent and its main route is the oxidative pentose phosphate pathway (Fig. 3–1). Glucose-6-phosphate is oxidized and decarboxylated in two steps to ribulose-5-phosphate. The reactions are catalyzed by glucose-6-phosphate dehydrogenase and 6-phospho-gluconate dehydrogenase, respectively. Both enzymes are present in high concentrations in blue-greens and both are NADP-specific: two molecules of NADPH are generated. The subsequent oxidation of NADPH during respiratory electron transport yields two molecules of ATP. The oxidative pentose phosphate pathway is activated in dark and inhibited in the light. The exact mechanism of this light inhibition is not yet fully understood.

Blue-greens are distinguished from other prokaryotes by their generally low rates of endogenous respiration and by their limited ability to utilize organic substances as a source of carbon and energy in the dark (see further). The reasons for this disparity are not quite clear and apparently complex. One is the finding that blue-greens, unlike aerobic heterotrophic bacteria and eukaryotic organisms, do not possess a complete tricarboxylic acid cycle (TCA or Krebs cycle). Some of the essential TCA cycle enzymes, like α-oxo-glutarate dehydrogenase, succinyl-CoA-synthase and succinic dehydrogenase, are present only in extremely low concentrations or absent from their cells (Fig. 3–3). (The situation is similar in other groups of autotrophic prokaryotes.) This

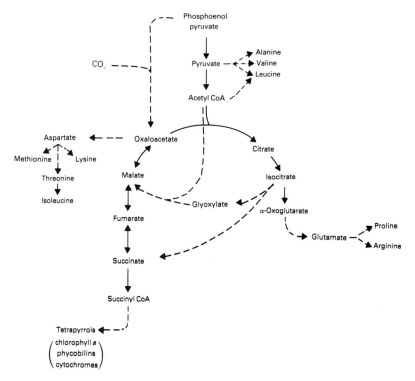

Fig. 3–3 The incomplete tricarboxylic acid cycle with the associated glyoxylate shunt and biosynthetic pathways in blue-greens (after Doolittle, 1979, *Adv. Microb. Physiol.*, **20**, 1–102).

deficiency prevents the cyclic flow of TCA cycle intermediates and consequently the utilization of substrates, such as acetate, pyruvate or tricarboxylic acids, for efficient energy production and growth in the dark. Though the incomplete TCA cycle does not function in substrate oxidation, it nevertheless performs a biosynthetic function enabling the synthesis of various amino acids and of lipids. Furthermore, enzymes of the glyoxylate shunt are present in blue-greens though their levels are relatively low (Fig. 3–3). This pathway permits a limited flow of carbon from isocitrate to succinate and further toward the synthesis of porphyrins like chlorophyll, cytochromes and phycobiliproteins.

The principal components of the respiratory electron transport chain appear to be present in blue-greens but there is so far little information concerning the flow of electrons between NADPH and O_2. It is, however, well documented that dark ATP synthesis is O_2 dependent (oxidative phosphorylation) and linked to the respiratory electron flow (Fig. 3–1).

3.3 Interrelations between photosynthesis and respiration

The rate of endogenous respiration in eukaryotic algae and in higher plants is in general similar in the light or in the dark. In blue-greens, however, respiratory O_2 uptake is affected by light: light of relatively low quantum dose (intensity) inhibits respiratory O_2 consumption, whereas at high quantum dose O_2 uptake is markedly increased. The interactions between photosynthesis and respiration are complex and to a great extent arise because of the lack of segregation of the two principal metabolic processes in prokaryotic cells. In contrast to the cellular organization in eukaryotic cells, photosynthetic and respiratory electron transport are closely linked in blue-greens and probably share many of the electron carriers.

Enhanced O_2 uptake in light of medium or high quantum dose, a phenomenon called photorespiration, is not restricted to blue-greens. It occurs also in eukaryotic algae and in higher plants. High O_2 tensions and low CO_2 concentrations enhance photorespiration. While dark endogenous respiration is saturated at about 5 kPa O_2, photorespiration increases linearly with increasing O_2 concentrations up to the atmospheric O_2 tension (20 kPa). Under these conditions the key photosynthetic enzyme, RuBP carboxylase functions as an oxygenase as well as carboxylase, effecting the cleavage and oxidation of RuBP to one molecule of 3-phosphoglycerate and one molecule phosphoglycollate. The associated inhibition of CO_2-fixation at high O_2 concentrations, occurs also in other prokaryotic and eukaryotic autotrophs, and was first detected by Warburg in 1920; it is known as Warburg effect. Phosphoglycollate gives rise in turn to glycollate, which is excreted by the cells of some blue-greens or oxidized to glyoxylate in other strains.

3.4 The utilization of organic substrates

Blue-greens were always regarded to be typical photosynthetic micro-organisms in which the light-dependent fixation of CO_2 is the dominant mode of nutrition. This concept, however, was somewhat obscured by the observation that in nature blue-greens are most abundant in habitats rich in organic matter, such as polluted lakes, shallow bodies of water, water-logged paddy fields or mangroves. Early reports on the ability of certain blue-greens to grow in the dark at the expense of organic substrates were viewed with some scepticism or at best were considered to be exceptions from the general rule of obligately photoautotrophic nutrition in blue-greens.

More recently it has become evident that the presence of organic substances in the growth medium can effect the metabolism and growth of blue-greens in a number of ways. Some organic compounds are not taken up but others assimilated and utilized as a source of carbon and/or energy by the cells. Some organic substances inhibit metabolic activities or may be highly toxic; others may stimulate growth without being assimilated.

Chlorogloea fritschii, a species common in cultivated soils of India, grows best in the dark in the presence of sucrose, and reasonably well with maltose. A few other organic compounds can support slow growth in the dark but many others fail to do so. Sucrose assimilation is more vigorous in the light, particularly in the absence of CO_2, and may result in a considerable increase in the growth rate.

On the basis of their response to organic substrates, blue-greens can be grouped into three main nutritional categories:

(i) Obligate photoautotrophs are organisms which depend entirely on photosynthesis; light is their sole source of energy and CO_2 is the principal source of carbon in their metabolism.

(ii) Facultative photoheterotrophs are able in the light to utilize certain organic substrates as a soure of carbon, in addition to or instead of CO_2, for the biosynthesis of cell material; they remain dependent on light reactions for the generation of high energy compounds (ATP) and reductant (NADPH).

(iii) Facultative chemoheterotrophs are capable of using organic substrates both as a source of carbon and energy in a light independent dark metabolism and for continuous slow growth in the dark; their heterotrophic potential is, however, increased in the light.

There appear to be several reasons for this great diversity in the heterotrophic potential among blue-greens. The primary requirement for the utilization of an organic compound is its transport across the cell membrane and its entry into the cell. The permeability of this membrane varies according to the organic substance, and may depend on the presence of specific carriers (permeases) which mediate the uptake of a particular substance. In the absence of such transport mechanisms substrates may still enter the cell by means of passive diffusion along a cencentration gradient. The rate of diffusion, however, is generally inadequate to support energy metabolism and growth in the dark.

Light invariably promotes the assimilation of organic substrates by blue-greens (Fig. 3–4), probably by providing light-generated ATP and/or reductant required partly for the active transport of solutes across the membrane and subsequently for the metabolism of organic substances incorporated. Photoassimilation of organic compounds may be increased in the absence of CO_2 or when CO_2-fixation is inhibited. The maximum rate of organic carbon assimilation, however, does not exceed the maximum rate of inorganic carbon (CO_2) fixation at saturating quantum doses of light: a clear indication of the dominant role of photosynthesis in the metabolism of blue-greens. Some organic acids (like acetic acid) and amino acids are assimilated even by obligately autotrophic species in the light, but their metabolism is limited to lipid and amino acid synthesis, and they can provide only a minor fraction (up to 10%) of the cells requirement for carbon.

Certain strains (like a *Nostoc* isolated from the cycad *Macrozamia*) are able to assimilate organic substrates immediately upon transfer in the dark. Others (like *Chlorogloea fritschii* and *Plectonema boryanum*) require a period of

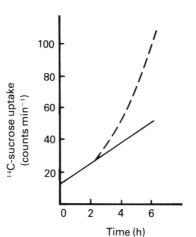

Fig. 3–4 Sucrose assimilation by *Chlorogloea fritschii* in light (– – – –) and in the dark (————) (after Evans and Carr, 1974, *Proc. 3rd Internat. Congr. Photosynth.*, 1861–5).

adaptation which may be associated with the synthesis of enzymes involved in dark heterotrophic metabolism. The lag is also attributable to the selection of a heterotrophic genotype. Appreciable dark heterotrophic growth is supported only by sugars, particularly glucose, sucrose and fructose. As under photosynthetic growth conditions, the principal route of carbohydrate metabolism in the dark is the oxidative pentose phosphate pathway.

3.5 Control of gene expression

A number of findings seemed to suggest that the low rate of growth in the dark may be due to a lack of effective regulation of enzyme synthesis at the transcriptional level and to a general regulatory inflexibility in blue-greens. Although growth of *Chlorogloea fritschii* in the dark may lead to a marked loss of phycocyanin, to a decreased chlorophyll content and also to greatly reduced photosynthetic activity, nevertheless the principal components of the photosynthetic apparatus are reproduced in the dark. Photosystem II activity may be lost but Photosystem I activity is retained during growth in the dark. The synthesis of RuBP carboxylase is not repressed when sugars are assimilated in the dark, and the entire Calvin cycle appears to remain functional in the dark, although at a greatly reduced rate. On the other hand, the concentration of enzymes concerned with the utilization of exogenous substrates, like amino acids and sugars, are either not increased at all or show only a small increase in the presence of appropriate substrates. This failure to adequately adjust the levels of enzymes involved in the intermediary metabolism has been suggested as one reason for obligate photoautotrophy of blue-greens. They appear to waste a great deal of energy through the synthesis of enzymes when they are not required. However, what appears to be wasteful may become an ecological advantage when environmental circumstances change; they are equipped to recommence their photoautrophic way of nutrition almost without delay upon re-exposure to light. Indeed, we may say that it would be in evolutionary terms irrational to abandon the photosynthetic mechanism in an environment where utilizable exogenous organic substrates are scarce or absent. It is more surprising that the reluctance by blue-greens to give up their metabolic independence is maintained even in situations where the supply of organic substrates is organizationally guaranteed, like in the symbiotic associations between blue-greens and higher plants (see Chapter 5).

Chlorogloea fritschii grown in the dark is able to perform photosynthesis immediately upon re-exposure to light, and the recovery of full photosynthetic capacity is accomplished within a few hours of incubation in the light. Both pigments and Calvin cycle enzymes are synthesized during the period of dark–light transition. Notably, sugar uptake is unaltered in the initial few hours in the light but thereafter it increases rapidly to the high rates usually observed when the organism is grown in the light. This may suggest the involvment of light-generated ATP in the active transport of sugars.

The negative response of many species to exogenous organic substrates gave rise to the misinterpretation that blue-greens in general are incapable of regulating their metabolism through repression and de-repression of enzyme synthesis. However, it is well documented that the synthesis of several enzymes (like nitrate reductase, nitrogenase, alkaline phosphatase or glucose-6-phosphate dehydrogenase) is controlled by induction and repression. It has also been shown that several polypeptides and enzymes apparently involved in dark energy metabolism (e.g. glycogen phosphorylase and glucose-6-phosphate dehydrogenase) are only synthesized by *Anacystis nidulans* following its transfer from light to dark growth conditions. Also, the formation of gas vesicle proteins in a *Nostoc* has been induced in the dark. Hence control of gene expression is as valid for blue-greens as for heterotrophic bacteria. Blue-greens, however, have evidently developed during their evolutionary history a different pattern of metabolic regulation which is relevant to their photoautotrophic mode of nutrition. This enables blue-greens to respond to the most important environmental stimuli, like light, CO_2 and other essential inorganic nutrients, and less effectively to organic substrates.

3.6 Nitrogen metabolism

Most blue-greens can utilize various forms of nitrogen for growth. Among the inorganic sources of nitrogen, ammonia is most readily incorporated into cells. Its assimilation proceeds via the glutamine synthetase-glutamate synthase (glutamine oxoglutarate aminotransferase, GOGAT) pathway:

$$NH_4^+ + \text{glutamate} + ATP \xrightarrow[\text{synthetase}]{\text{glutamine}} \text{glutamine} + ADP + P_i + H_2O \quad (1)$$

$$\text{glutamine} + \alpha\text{-oxoglutarate} + NADPH \xrightarrow{\text{GOGAT}} 2\,\text{glutamate} + NADP \quad (2)$$

Other ammonia assimilating routes may operate at higher ammonia concentrations. Enzymes involved in these pathways (glutamate dehydrogenase, alanine dehydrogenase or aspartate dehydrogenase) have also been detected in extracts from several blue-greens.

Nitrate is the commonest source of combined nitrogen for blue-greens in nature. Its assimilation follows a sequence of reductions to ammonia, catalyzed by the enzymes nitrate reductase and nitrite reductase, similar to those established for green algae and higher plants:

$$NO_3^- \xrightarrow[\text{reductase}]{\text{nitrate}} NO_2^- \xrightarrow[\text{reductase}]{\text{nitrite}} NH_4^+$$

Ferredoxin acts as the physiological electron donor in the two reactions. Both nitrate reductase and nitrite reductase are induced by their respective substrates, although the additional presence of ammonia represses the synthesis of nitrate reductase.

Among the organic sources of nitrogen, urea has been found to support growth of certain blue-greens in culture. Various amino acids, amides and amino acid mixtures can serve as a source of nitrogen for other strains. A few amino acids can act as a sole source of carbon and nitrogen for the slow growth in the dark of certain blue-greens.

3.7 Nitrogen fixation

Biological N_2-fixation is the reduction of elemental nitrogen to ammonia, catalyzed by the complex enzyme system nitrogenase:

$$N_2 + 6H^+ + 6e^- \xrightarrow{\text{nitrogenase}} 2NH_3$$

This is one of the fundamental biological processes, essential for the maintenance of the nitrogen status of the whole biosphere. N_2-fixation takes place only in certain prokaryotic micro-organisms that are able to synthesize nitrogenase; they are either free living or in symbiotic associations with other organisms (see Chapter 5). The reaction requires a low potential reductant (ferredoxin or flavodoxin), magnesium ions (Mg^{2+}) and a considerable input of metabolic energy in the form of ATP. 12 to 15 molecules of ATP are consumed for each molecule of N_2 reduced. The high energy cost of N_2-fixation may explain why N_2-fixing organisms have developed devises for the repression of nitrogenase synthesis when energetically less expensive sources of combined nitrogen are present in their environment.

Nitrogenase is extremely sensitive to free O_2 and can function only under anaerobic conditions. Direct exposure of the enzyme to air results in the inactivation and even the destruction of the component proteins. The O_2 sensitivity of nitrogenase and the fact that the enzyme system is present only in prokaryotes suggest that the ability to fix N_2 has originally evolved during the early anoxygenic period of Earth's history. There is reasonable geochemical and fossil evidence in support of this suggestion. With the development, initially in ancient blue-greens, of O_2-evolving photosynthesis, the atmosphere has become gradually more oxygenic until the partial pressure of O_2 in the atmosphere reached its present day level of 20 kPa (see also section 6.3). This profound change acted as a selective evolutionary force upon N_2-fixing micro-organisms, including blue-greens. Some have become restricted in their distribution and are confined to the remaining anaerobic or microaerobic habitats. Those N_2-fixing forms that have become adapted to an oxygenic environment have had to acquire adequate devices for the protection of nitrogenase against inactivation by O_2. In addition blue-greens had a special 'task' to protect their nitrogenase from the O_2 liberated as a result of their own photosynthetic activities. Far the most effective adaptation for the protection of nitrogenase in blue-greens was the acquisition of a highly specialized N_2-fixation cell, the heterocyst (see section 2.1). Heterocystous blue-greens are able to fix N_2 efficiently under ambient atmospheric conditions. Some non-heterocystous forms have apparently succeeded in maintaining their ability to synthesize nitrogenase and to fix N_2 in the absence or at low concentrations of exogenous O_2 (Table 3). Evidently, anaerobic conditions are very difficult to maintain in cells that evolve O_2 in the course of their primary metabolism.

Table 3   Non-heterocystous blue-greens able to fix N_2 (after Stewart 1980, *Ann. Rev. Microbiol.*, **34**, 497–536).

Section	Order/Family	Genus	No. of strains tested	No. of N_2-fixing strains
I	Chroococcales	*Gloeobacter*	1	0
		Gloeothece	5	5
		Synechococcus	26	5
		Gloeocapsa	4	0
		Synechocystis	16	0
II	Chamaesiphonales	*Dermocarpa*	6	2
		Dermocarpella	1	0
		Chroococcidiopsis	8	8
	Pleurocapsales	*Myxosarcina*	2	1
		Xenococcus	3	1
		Pleurocapsa	12	7
III	Oscillatoriaceae	*Spirulina*	2	0
		Oscillatoria	9	5
		Lyngbya *Phormidium* group *Plectonema*	24	16
		Pseudanabaena	8	3
		Total	127	53 (41.7%)

N$_2$-fixing blue-greens have the simplest nutritional requirements of all living organisms. They need only N$_2$, CO$_2$, water and mineral elements for growth in the light, utilizing solar energy to drive their metabolic and biosynthetic machinery. However, the extent to which blue-greens can profit from their remarkable autotrophic potential depends on their ability to reconcile the presence of an O$_2$-sensitive nitrogenase side by side with an O$_2$-evolving photosynthetic apparatus. It is noteworthy that in general blue-greens are more sensitive to high O$_2$ concentrations and to high quantum doses of light (which could result in increased O$_2$ evolution) when they are grown under N$_2$-fixing conditions (Fig. 3–5).

Heterocysts provide an ideally suited microenvironment for N$_2$-fixation in that they maintain the reducing conditions essential for nitrogenase activity. This results from a combination of several factors, like the absence of O$_2$-evolving photosynthesis, an active oxidative metabolism and the presence of an elaborate envelope which reduces the diffusion of atmospheric gases through the surface of the heterocyst to a level which is adequate to saturate nitrogenase with the substrate N$_2$; the O$_2$ which seeps in can be dealt with by means of respiration. The protection of nitrogenase against oxygen damage is further enforced by nitrogenase catalysed H$_2$ evolution. H$_2$ is a by-product of the nitrogenase reaction, and accounts for about 25 to 30% of the total electron flux by nitrogenase. H$_2$ may help to exclude O$_2$ from the site of nitrogenase; furthermore the oxidation of H$_2$ in the so called oxyhydrogen (or Knallgas) reaction, catalyzed by an 'uptake hydrogenase', may contribute significantly to the removal of free O$_2$ from the heterocysts.

Heterocysts possess adequate mechanisms to meet the requirements of N$_2$-fixation for reductant and energy. By retaining an active Photosystem I,

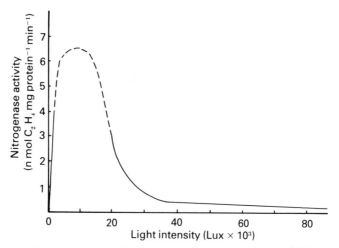

Fig. 3–5 Nitrogenase activity (acetylene reduction) as a function of light quantity (after Reynaud and Roger, 1979, *C.R. Acad. Sci. Paris Ser. D.*, **288**, 999–1002).

they are capable of generating reducing power and ATP in cyclic photo-electron transport coupled with photophosphorylation. They are also able to provide the same requirements, though less efficiently, in an O_2-dependent dark respiratory metabolism.

A consequence of the inability of heterocysts to perform photosynthetic CO_2-fixation is that they must rely on the vegetative cells for a supply of carbon and a source of reductant. Carbon (possibly in the form of a disaccharide sugar) is translocated presumably through the fine channels (microplasmodesmata) traversing the small septum between vegetative cells and heterocysts. Fixed nitrogen (probably in the form of glutamine) may be exported through the same routes but in an opposite direction from the heterocysts to vegetative cells (Fig. 3–6). The accumulation of cyanophycin polypeptide and the deposition of 'plug' material in the heterocyst (see sections 1.6 and 2.1) appear to have an important role in the regulation of supply of fixed nitrogen to vegetative cells. Enzymes involved in both the synthesis and degradation of cyanophycin are several times more active in heterocysts than in vegetative cells. Cyanophycin in heterocysts may function as reservoir between N_2-fixation and the export of fixed nitrogen.

Among the N_2-fixing non-heterocystous blue-greens one can observe a variety of apparently less efficient and probably more ancient adaptations of the N_2-fixing system to the prevailing oxygenic environment which enable N_2-fixation to proceed under microaerobic conditions.

The first discovery of N_2-fixing in a non-heterocystous species was made in 1961 with bloom-forming planktonic populations of the marine *Oscillatoria* (*Trichodesmium*) *erythraea*, common in tropical oceans. The trichomes are mostly aggregated in bundles (Fig. 3–7), and N_2-fixing activity is associated with this typical bundled morphology. The individual trichomes within the bundle display a distinct morphological and physiological differentiation through the segregation of photosynthetic and non-photosynthetic regions. The former appear granulated, contain carboxysomes and have been shown to incorporate ^{14}C-labelled CO_2. The latter lack these features but show strong reducing potential, harbor attached bacteria and seem to provide favourable environment for the maintenance of nitrogenase. It is thought that O_2 tensions in the bundle centre could become sufficiently decreased to permit N_2-fixation taking place in the non-photosynthesizing cells. Thus in *O. erthraea* the contrasting physiological activities are restricted to separate regions of the trichome.

Another case of aerobic N_2-fixation is that carried out by a few closely related (possibly identical) strains of the pleurocapsalean genus *Gloeothece* (*Gloeocapsa*), which display a typical colonial morphology. The cell aggregate is surrounded by a lamellate envelope which encloses two or three cell generations produced in subsequent cell divisions (Fig. 1–1d). Although *Gloeothece* can fix N_2 in air, optimum rates of nitrogenase activity have been observed under subatmospheric O_2 tension (10 kPa), at low irradiance, and low CO_2 concentrations. All these seem to indicate that the organism is relatively

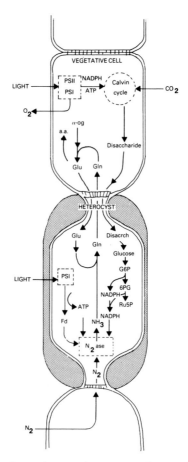

Fig. 3–6 Metabolic exchange between heterocyst and vegetative cell. PSI, Photosystem I; PSII, Photosystem II; a.a., amino acids; α og, α-oxoglutarate; Glu, glutamate; Gln, glutamine, G 6P, glucose-6-phosphate; 6PG, 6-phosphogluconate; Ru5P, ribulose-5-phosphate; Fd, ferredoxin; N₂ase, nitrogenase.

O_2-sensitive under N_2-fixing conditions. Several suggestions have been put forward to explain how nitrogenase might be protected in *Gloeothece* against inactivation by O_2. Some refer to the protective role of the elaborate envelope, other to the possibility of a temporal or spatial segregation of photosynthetic and N_2-fixing activities.

There are a few other, not fully confirmed, reports of 'aerobic' N_2-fixation by non-heterocystous blue-greens, but many more records of nitrogenase synthesis and N_2-fixation in non-heterocystous forms while maintained under strict anaerobic conditions, as provided by continuous sparging of cultures with an O_2-free gas mixture ($N_2 + CO_2$). This was first demonstrated in *Plectonema*

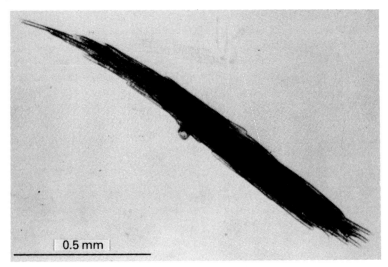

Fig. 3–7 Filament bundle of *Oscillatoria* (*Trichodesmium*) (light micrograph). Courtesy of Dr I. Bryceson.

boryanum and later in several other non-heterocystous strains (see Table 3). The latent ability (genetic information) to fix N_2 is expressed only in the absence of free O_2. It is concievable that such control by O_2 of nitrogenase synthesis is beneficial in ecological terms, as it prevents the wasteful process of nitrogenase synthesis under undesirable conditions, when the enzyme produced would only be doomed for destruction.

3.8 Growth characteristics

Bacteriologists and molecular biologists, who are used to working with fast-growing bacteria like *E. coli*, which divide every 20 minutes in a nutrient rich medium, are often discouraged by the relatively slow growth of blue-greens. Under favourable conditions (saturating illumination, optimum temperature and adequate supply of nutrients) the generaion time of most blue-greens varies between about 12 and 36 hours. Such a growth rate could be an adaptation to the natural diel cycle (24 h cycle) and their predominantly photoautotrophic metabolism. But there are also some exceptions. A recently isolated marine *Anabaena* has a generation time of only 4.3 hours (which corresponds to 5.6 generations per day) when grown under optimum conditions. The fast growing of all blue-greens so far isolated in culture is a thermophilic *Synechococcus* (*Anacystis nidulans*); it has a doubling time of only 2 hours (12 generations per day) when grown under optimum conditions (41°C). Growth rates and generation times of some blue-greens extensively used in research are given in Table 4.

Growth rates of blue-greens are not increased when more reduced sources of carbon and nitrogen (like sugars and ammonia, respectively) are provided in the medium in addition to or in place of CO_2, nitrate or N_2, respectively. Growth rates during heterotrophic growth in the dark are in general much lower than when the same organisms are grown in the light in the presence or absence of an organic substrate. The reasons for the slow growth in the dark are probably complex and not fully understood. The incomplete tricarboxylic acid cycle may be responsible for the relatively inefficient oxidative metabolism of carbohydrates. Rates of oxidative phosphorylation are certainly several times lower than rates of photosynthetic phosphorylation in blue-greens.

Table 4   Specific growth rates and generation times of some blue-greens.

Strain	Specific growth rate (k)*	Generation time (h)
Anabaena cylindrica	0.51	14.1
Anabaena strain CA	1.68	4.3
Agmenellum quadruplicatum	–	2.15
Synechococcus (*Anacystis nidulans*)	0.87	2.03

*Specific growth rates are expressed in $\log_{10}$ units per day; when $k = 0.301$, the generation time is 24 h.

4 Ecology

Blue-greens are common in all kinds of natural habitats. Many species are cosmopolitan, distributed throughout the world. Certain blue-greens are more or less permanent and often dominant components of the plankton of tropical, temperate and frigid lakes and oceans, others are characteristic of streams, rivers, hot springs or rocky seashores. Some are well adapted to terrestrial habitats, and are particularly abundant in tropical soils and in waterlogged fields. Blue-greens may be present in extremely hostile and most peculiar habitats. They are known to be amongst the earliest colonizers of arid land. Only eighteen months after a volcanic eruption on the island of Heimaey, near Iceland, *Anabaena* was found growing and fixing N_2 on the lava soil. It would, however, be wrong to give the impression that blue-greens are a universally successful group of micro-organisms. There are many habitats, like for example acid streams and lakes, where blue-greens are rather uncommon or totally absent.

4.1 Blue-greens in freshwater: planktonic forms

Unicellular and filamentous blue-greens are almost invariably present in freshwater lakes frequently forming dense planktonic populations or water blooms in eutrophic (nutrient rich) waters. In tropical regions growth of blue-green populations may be continuous throughout the year, but in temperate lakes there is a characteristic seasonal succession of the bloom-forming species, due apparently to their differing responses to the physical-chemical conditions created by thermal stratification. Usually the filamentous forms (such as various *Anabaena* species, *Aphanizomenon flos-aquae* and *Gloeotrichia echinulata*) develop first soon after the onset of stratification in late spring or early summer, while the unicellular-colonial forms (like *Microcystis* species) typically bloom in mid-summer or in autumn. The main factors which appear to determine the development of planktonic populations are light, temperature, pH, nutrient concentrations and the presence of organic solutes.

Many planktonic blue-greens grow best at relatively low quantum doses of light (low light intensities or photon flux densities) which may be sufficient to saturate their photosynthetic metabolism. This provides blue-greens with a distinct competitive advantage as demonstrated in a mixed chemostat culture of *Oscillatoria* with the green unicellular alga *Scenedesmus*. Although under optimum conditions *Scenedesmus* can grow twice as fast as *Oscillatoria*, the range of photon flux density (between 30 and 85 W m^{-2}) which allows maximum growth of *Scenedesmus* is more than three times higher than that required for maximum growth of *Oscillatoria* (6 to 25 W m^{-2}). Photon flux density which inhibits growth (25 W m^{-2}) is also much lower for *Oscillatoria* than for *Scenedesmus* (85 W m^{-2}). *Oscillatoria* is able to grow at very low quantum doses of light (below 2 W m^{-2}) efficiently harvesting the available light energy for growth, whereas *Scenedesmus* cannot compensate for its maintenance at this low level of illumination.

Temperature is probably a less important factor in the growth of blue-green populations than generally assumed. Insufficient light may not be primarily responsible for the absence of blue-greens in the early spring period. Although growth and photosynthesis is optimal around 20–30°C, signficant photosynthesis takes place at much lower temperatures. A more important factor appears to be the pH of the lake water; most forms grow best in the neutral pH region and tolerate better alkaline than acid conditions. This is partly due to their ability to scavenge CO_2 efficiently even at very low concentrations in water and also to utilize bicarbonate ions as a source of carbon in photosynthesis (see § 3.1).

Nutrient concentrations may be critical for the development of planktonic populations. Prolonged deficiency of essential nutrients, particularly phosphorus and nitrogen, will inevitably limit their growth. Their ability to accumulate, by 'luxury consumption' when such nutrients are available in their environment, large concentrated reserves of these substances makes blue-greens less dependent on the fluctuating supply of nutrients. Furthermore, heterocystous forms are able to grow by utilizing atmospheric N_2 when combined forms of nitrogen are not available. Significant N_2-fixation in temperate and tropical lakes has been shown to be associated with the occurrence of heterocystous populations (Fig. 4.1).

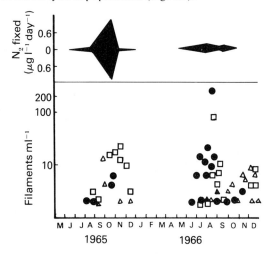

Fig. 4–1 Variation in numbers of planktonic blue-greens in relation to nitrogen fixation in Windermere south basin. ●, *Anabaena flos-aquae*; □, *A. solitaria*; ▲, *A. circinalis*; △, *Apanizomenon flos-aquae* (after Horne and Fogg, 1970, *Proc. R. Soc. B.*, **175**, 351–66).

The correlation between the concentration of organic solutes and the abundance of planktonic blue-greens has been mentioned in section 3.4. Facultative heterotrophy appears to be widespread amongst planktonic blue-greens, and could contribute to the maintenance of populations during conditions of limited photosynthesis. Heterotrophic bacteria, however, are incomparably more efficient in the utilization of organic substances. In any case, the oxidative metabolism of organic substrates will result in decreased amounts of dissolved O_2. Partial deoxygenation of lake water will in turn stimulate both photosynthesis and N_2-fixation, and thus the development of planktonic blue-green populations.

4.1.1 Buoyancy regulation

Light is indispensible for blue-greens but photon flux density attenuates exponentially with depth in water. It is therefore advantageous that planktonic populations maintain their position in water at an optimum depth for photosynthesis and growth. The tiny cylindrical gas vesicles confer buoyancy to planktonic blue-greens (see § 1.7) but the buoyancy needs to be finely regulated to acquire the desirable vertical position: not too close to the surface of the water, where the intensity of solar radiation could be detrimental, and not too far below the surface where the available illumination is too weak to support photosynthetic growth.

It has been shown that there is an inverse relationship between irradiance and gas vacuolation in the cells of *Anabaena flos-aquae*. When grown in dim light, the filaments become more buoyant, due to an increased rate of gas vesicle formation. In contrast, buoyancy is lost and the filaments gradually sink when cultures are exposed to intense radiation. Turgor pressure is higher in cells exposed to strong light than in cells under weak illumination, and the increase in turgor pressure causes the collapse of gas vesicles. The rise in turgor pressure is partly the result of increased photosynthetic activity accompanied by increased production of sugars and other small organic molecules, and partly due to light-stimulated uptake of potassium ions. The collapse of a proportion of gas vesicles can cause the cells to lose their buoyancy.

The two counteracting mechanisms, which stimulate gas vesicle formation in dim light and effect the partial collapse of gas vesicles in strong light, enable the organisms to regulate their buoyancy. The formation of excess gas vesicles at decreased irradiance causes the rise of planktonic populations to the upper well-illuminated (euphotic) zone, whereas the collapse of gas vesicles prevents the accumulation of filaments at the surface where solar radiation could be of lethal density. Hence the population stratifies on a vertical light gradient: the blue-greens float up as new gas vesicles form in dim light, and sink down when increased photosynthesis leads to the partial collapse of gas vesicles (see Fig. 4–2).

Those planktonic forms, like *Aphanizomenon flos-aquae*, which tend to form aggregates, move faster through the water column than single cells or single filaments. They may display on calm summer days a diurnal vertical migration regulated by similar mechanisms.

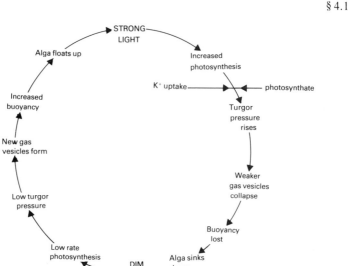

Fig. 4–2 Diagram showing the sequence of buoyancy regulation in planktonic blue-greens (after Reynold and Walsby, 1975, *Biol. Rev.*, **50**, 437–81).

4.1.2 Water blooms

The formation of water blooms results from the redistribution and often rapid accumulation of buoyant planktonic populations (Fig. 4–3). When such populations are subjected to suboptimal conditions, they respond by increasing their buoyancy and move upward nearer to the water surface. Water turbulence usually prevents them reaching the surface. If, however, turbulence suddenly weakens on a calm summer day, the buoyant population may 'over-float' and rise very close to the surface, or indeed become lodged right at the water surface (Fig. 4–4). There the cells are exposed to most unfavourable and dangerous conditions, like O_2 supersaturation, rapidly diminishing CO_2 concentrations and intense solar radiation, which are inhibitory to photosynthesis and N_2-fixation, causing photo-oxidation of pigments and

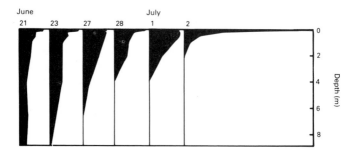

Fig. 4–3 Changes in the vertical distribution of *Anabaena circinalis* in Crose Mere, Shropshire, during 1968 (after Reynold, 1971, *Fld Stud.*, **3**, 409–32).

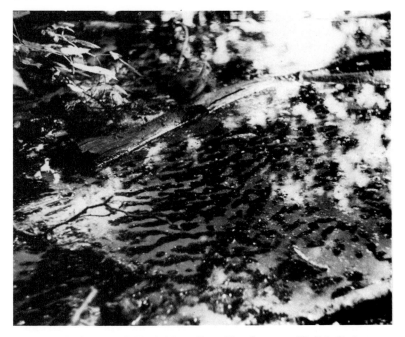

Fig. 4–4 Surface scum in Kettle Mere, Shropshire. Courtesy of Dr J.A. Rother.

inflicting irreversible damage to the genetic constitution of cells. A frequent outcome of surface bloom formation is massive cell lysis and rapid disintegration of large planktonic populations. This is closely followed by an equally rapid increase in bacterial numbers, and in turn by a fast deoxygenation of surface waters which could be detrimental to animal populations within the lake. Water blooms are objectionable for recreational activities, and more importantly, create great nuisance in the management of water reservoirs.

An important adaptation of heterocystous species to the hazardous situation created by increased population density during a water bloom is the formation of akinetes (see § 2.5). Their production coincides with the stabilization of the water column and often proceeds rapidly and extensively in apparent response to conditions unfavourable for vegetative growth. Protected by their thick capsule, akinetes can withstand the hostile environmental conditions at the water surface. While the vegetative cell material disintegrates, the detached akinetes, which lack gas vesicles, will slowly descend through the water column. On their way down and while passing through layers containing adequate supplies of nutrients, a proportion of the sinking akinetes may germinate thus giving rise to new vegetative material before they would pass out of the photic zone. The germlings and juvenile filaments constitute the nucleus of a new planktonic population which usually remains in suspension throughout the winter period and forms the inoculum for spring growth. The rest of the akinetes sediments to the mud surface. A fraction of these akinetes may successfully overwinter there, and may be brought back into suspension the following spring by water mixing and turbulence. Reaching the upper layers of water they may germinate in the light and contribute to the build-up of planktonic material. Thus akinetes help to preserve the species both during catastrophic water blooms and following the harsh winter period in temperate lakes.

Apart from bloom formation, a proportion of the planktonic blue-green population steadily sinks, decomposes and sediments at the lake bottom. The standing crop rapidly declines in the autumn leaving only a small residual population floating in the water or overlying the bottom sediment. Non-heterocystous forms, like *Microcystis*, overwinter there in vegetative form without being able to produce spores.

4.2 Attached and benthic populations in lakes

Many blue-greens grow attached on the surface of rocks and stones (epilithic forms), on submerged plants (epiphytic forms) or on the bottom sediments (epipelic forms) of lakes. The latter community is also referred to as benthos. These associations are not well defined due to their irregular distribution and complexity which renders an analytical or physiological investigation rather difficult. In very shallow lakes the attached forms could make a significant contribution to the productivity of the lake.

The epilithic community displays a clearly discernable zonation in lakes. The composition of blue-green populations around the water line may differ markedly from those above the water level (spray zone) and below the surface. The zonation is apparently determined by the ability of species to withstand prolonged exposures to strong light and to desiccation. Members of the genera *Pleurocapsa*, *Gloeocapsa* and *Phormidium* are well adapted to these habitats and often dominate the dark blue-black community of the spray zone. *Scytonema* and *Nostoc* species form olive-green coatings and are more frequent about the water line, whilst the brownish *Tolypothrix* and *Calothrix* species are more typical components of the subsurface littoral community. The varied coloured appearance of the component species is partly due to the pigmentation of their sheaths but may also be the result of chromatic adaptation.

The epiphytic flora of lakes is usually dominated by diatoms and green algae, and blue-greens are of less importance in this community. Species of the genera *Nostoc*, *Lyngbya*, *Chamaesiphon* and *Gloeotrichia* have been found occasionally encrusting submerged plants. The loose epipelic community, which spreads over the bottom sediment, commonly includes blue-greens (like *Aphanothece* and *Nostoc*) particularly in the more eutrophic lakes. Benthic blue-greens growing over the littoral sediments and on submerged plants may be responsible for the occasional high rates of N_2-fixation measured in oligotrophic lakes.

4.3 Blue-greens in streams and rivers

For obvious reasons streams and rivers can maintain only a limited and continuously changing planktonic population since the floating forms are permanently swept downstream. Planktonic blue-greens are rather uncommon in rivers, except during periods of reduced flow. But attached forms, well adapted to strong currents, may grow abundantly receiving a continuous supply of nutrients. Many blue-greens are epiphytic on larger algae, like *Cladophora*, and on other river plants, including bryophytes and angiosperms.

The size of the epilithic flora in streams and rivers is to a large extent effected by the chemistry of the water and substratum. Calcareous rocks, for example, can support a larger and more complex community than non-calcareous rocks. The composition of the algal population is further influenced by light and the rate of current flow. *Chamaesiphon, Nostoc, Rivularia, Phormidium, Pleurocapsa* and *Homoethrix* species are among the more common blue-green inhabitants of this environment.

A group of blue-greens is associated with carbonate precipitation in calcareous springs. Carbonate deposition tends to bury the colonies, and they respond by creeping and growing over the crust layers. The process leads to the formation of porous carbonate sediments, such as travertine.

4.4 Blue-greens in thermal waters

Geothermal springs of temperatures between 45 and 100°C are almost exclusively inhabited by thermophilic prokaryotic micro-organisms, blue-green algae, photosynthetic flexibacteria, and certain non-photosynthetic autotrophic and heterotrophic bacteria. Thermophilic blue-greens were found to grow at tempratures as high as 73°C but non-photosynthetic autotrophic bacteria can tolerate even higher temperatures up to 95°C! Blue-greens are most abundant in the alkaline hot springs and absent from the hot acid springs. The hot source water contains most of the essential nutrients and permits a steady growth of blue-greens throughout the year.

As the effluent moves away from the source of the spring, the water cools thus setting up a temperature gradient, which steepens as it expands. It is accompanied by concentration gradients of sulphide (decreasing due to oxidation), O_2 (increasing) and pH (increasing), and by concomitant gradual changes in the composition of the microflora (Fig. 4–5). A typical succession of species in hot springs of Oregon begins with the filamentous flexibacterium *Chloroflexus* which forms thick orange-green mats at the source spring 70 to 90°C hot. Downstream as temperature gradually decreases from 73 to about 53°C populations of the unicellular blue-green *Synechococcus lividus* form deep green to yellow green mats about 1–2 mm thick. These are succeeded by the reddish-brown mats of the filamentous gliding *Oscillatoria terebriformis*. The last zone of the thermal stream (with temperatures between 35–45°C) is dominated by the association of *Pleurocapsa* and *Calothrix* species which form a dark leathery crust on the substratum. A similar pattern of zonation can be seen in alkaline and neutral hot springs in Yellowstone Park, Iceland or New Zealand, with some variations in species composition of the local flora.

Other less common thermal habitats attended by blue-greens are the meromictic (seldom-mixing) saline lakes (where the heavy saline bottom waters can reach a temperature of up to 55°C or more), the small freshwater

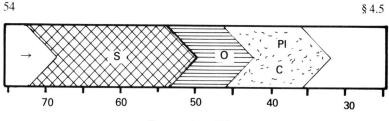

Temperature (°C)

Fig. 4–5 Species pattern in a stream of Hunter's Hot Springs, Oregon. →, direction of water flow; S, *Synechococcus*; O, *Oscillatoria*; Pl, *Plectonema*; C, *Calothrix* (after Castenholz, in CARR and WHITTON, 1973 *The Biology of Blue-Green Algae*. Blackwell, Oxford).

and saline pools in hot deserts, and the hot mountain cliffs covered by consipicuous ink-dark streaks of blue-greens. All these are heated by direct or indirect solar radiation.

The unique ability of prokaryotic micro-organisms to endure very high temperatures may be due to the inheritance of ancient characters suited to the environmental conditions which were prominent in the early periods of Earth's history. It has been shown that the thermophilic strains possess a highly stable photosynthetic apparatus; their temperature optima for photosynthesis is extremely high, similar or even higher than the temperature regime of water from which they were collected (Fig. 4–6). Also, thermophilic strains are extremely sensitive to low temperatures; sudden exposure to suboptimal temperatures causes a cold shock syndrome which leads to bleaching, lysis and cell death. It appears that cold-sensitivity is a membrane related phenomenon associated with a low content of polyunsaturated fatty acids of cellular membranes. Another reason for the dominant role of blue-greens in hot springs is most certainly the absence of competitors (eukaryotic algae) and of grazers, which are unable to tolerate the prevailing high temperatures.

4.5 Blue-greens in the marine environment

The presence of blue-greens is in general less evident in the open sea than in freshwaters, and it is thought that they are less important in terms of biomass and productivity in the marine environment. Nevertheless, blue-greens are of widespread distribution in and around the oceans and may form an important element of the vegetation of some particular marine habitats, like the intertidal zone of temperate and tropical seas and in estuarine areas. All major groups, with the exception of the Stigonematales, are well represented.

4.5.1 Planktonic and benthic forms

Until recently blue-greens were thought to be an insignificant part of the marine plankton as only a few truly planktonic forms had been recorded. In tropical waters the most important are the populations of the marine *Oscillatoria* (*Trichodesmium*) (see § 3.7) which frequently form extensive surface blooms in the shape of long orange-brown windrows (known to sailors as 'sea-sawdust') in the Red Sea, the Indian Ocean and in the equatorial regions of the Pacific

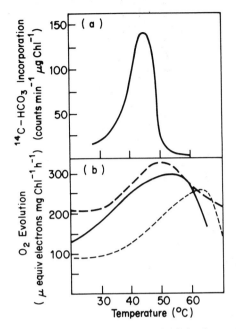

Fig. 4–6 Temperature dependence of photosynthesis in thermophilic blue-greens. (a) *Spirulina labyrinthiformis* (after Castenholz, 1977, *Microb. Ecol.*, **3**, 79–105). (b) *Mastigocladus laminosus*, grown at 40°C (———), 45°C (– – – –) and 55°C (........), respectively (after Bohler and Binder, 1980, *Arch. Microbiol.*, **124**, 155–160).

and Atlantic Oceans. The buoyant populations of *Oscillatoria* normally occupy a zone about 15–25 m below surface, though they may extend down to as deep as 175 m. Under calm water conditions the filament bundles (Fig. 3–7) ascend and accumulate near the water surface, where they may become trapped and undergo photo-oxidation and cell lysis. The phycoerythrin released from the lysing trichomes may colour the surrounding waters bright purple. High rates of N_2-fixation were measured in material collected from subsurface layers, and rates may be even higher in the deeper water where the physical conditions are more favourable for N_2-fixation. Thus the contribution of the marine *Oscillatoria* to the nitrogen budget of the open sea is probably formidable.

Reddish (phycoerythrin-rich) floating cells of a marine *Synechococcus* were recently found to be a universal component of the marine phytoplankton. It had remained undetected so long because of its extremely small size (1–2 μm), far below the mesh size of the finest plankton nets. They are most numerous (10^4 to 10^5 cells ml^{-1}) in the top 20 m of sea water. The ecological significance of this species is as yet uncertain but it may well be considerable judging from its ability to photosynthesize efficiently and to grow well at the relatively low photon flux densities present at the bottom of the euphotic zone.

Other planktonic blue-greens, which were recorded earlier from tropical and subtropical sea-water are of doubtful importance. Occasional occurrences of

Calothrix and *Anabaena* species near the coastal waters are most probably attributable to littoral material being washed out to the open sea.

Little is known of the distribution of blue-greens in the marine sublittoral zone. Sporadic records from tropical and subtropical regions suggest that benthic species (which belong mainly to Oscillatoriaceae) may be present down to a depth of 20–30 m below water level. Light may still penetrate sufficiently in clear waters to this depth to support photosynthetic growth.

4.5.2 Intertidal communities

Epilithic blue-greens are common constituents of the littoral fringe of temperate and tropical seas and form often conspicuous growths on rocky shores. They may appear as black encrusting sheaths or thick cushions over the soft limestone or sandstone rocks. Though the composition of this community varies with the season, the prominent forms belong to the genera *Calothrix*, *Gloeocapsa*, *Phormidium* and *Nodularia*. *Calothrix* is particularly important in Scottish shores and a related species appears to be a component (phycobiont) of the littoral lichens *Lichina* and *Verrucaria*, which are closely associated with the blue-green belt. Their N_2-fixing activities no doubt contribute considerably to the productivity of the littoral zone.

Lime-boring blue-greens often dominate the endolithic community which grows in the inner pore and cleft faces of limestone cliffs produced by the combined erosive action of atmospheric weathering and biological dissolution processes.

Filamentous blue-greens are abundant in the intertidal region of coral reefs. They constitute the primary source of food for the associated herbivores and play an important part in the high productivity of reef and tropical lagoon communities. Blue-greens have also been found to be abundant in the coastal mangrove forests. While the heterocystous forms grow and fix N_2 in aerobic habitats, the non-heterocystous populations develop in close contact with anaerobic sediments and also display nitrogenase activity.

Characteristic of salt marshes is the gradual transition from freshwater to marine habitats. Blue-greens in general show an exceptional ability to withstand great variations in the salinity of water which explains the universal presence of blue-green population in the estuarine environment. They are also well fitted to tolerate desiccation and considerable fluctuations in the redox potential and pH of the medium. The algal community of salt marshes, which also includes diatoms and green algae, displays a characteristic zonation. Beginning from the muddy banks and creeks and proceeding upward to the top of the marsh, the blue-green population is dominated in sequence by forms belonging to the genera *Oscillatoria* and *Phormidium*, then *Lyngbya*, further *Phormidium* and *Calothrix*, and finally *Rivularia* and *Nostoc*. Many filamentous strains may play an important part in stabilizing mud surfaces and in the formation of a humus layer.

Intertidal blue-greens are distinguished by their adaptability to both freshwater and marine environment. Freshwater isolates can grow in presence of greatly increased salt concentrations in the medium, and conversely, marine isolates can be maintained at salinities near to those characteristic of freshwaters. There are well known records of the occurrence and dominance of blue-greens in hypersaline ponds, like the Yallah Pond in Jamaica or the Solar Lake in the Sinai peninsula.

4.6 Terrestrial blue-greens

Although blue-greens are more apparent in aquatic habitats, they can often be seen on the surface of moist soil and growing beneath the soil surface. In the temperate region blue-greens are especially common in calcareous and alkaline soils. Certain species, like *Nostoc commune*, are often conspicuous on the soil surface. Acid soils, however, lack blue-green element and are usually dominated by diatoms and green algae.

Blue-greens are almost universal components of the flora in tropical soils. They show particularly rich growth in shallow pools and waterlogged fields. From soils of tropical Central and South America and the Caribbean Islands, 46 out of 62 species isolated were found to belong to blue-greens. *Schizothrix, Scytonema, Phormidium, Nostoc* and *Microcoleus* are of world-wide distribution in the microflora of subtropical and tropical arid soils. In areas where high temperature combines with high humidity, blue-greens will grow luxuriantly also on sub-aerial substrates, like rocks, the bark of trees or the walls and roofs of buildings. They are readily recognizable by the deep brown or blackish tufts and crusts produced by such populations.

Blue-greens are frequently encountered in the polar regions. They are major components of the soil microflora in the Antarctic and in the arctic tundra of Alaska. Heterocystous species (mainly *Nostoc*) have been found active in N_2-fixation. Free-living and epiphytic N_2-fixing strains are also common in the arctic meadows and peats.

Growth and propagation of terrestrial blue-greens is controlled by moisture, soil temperature, irradiance and pH. Soil desiccation brought about by drought and intense solar radiation is well tolerated by blue-greens probably because of the protection provided by the thick mucilage sheaths. Heterocystous forms respond to adverse conditions by extensive sporulation. The akinetes produced are able to withstand long periods of drought and may readily germinate upon rewetting of the soil.

Blue-greens improve soil fertility in a variety of ways. Many fix N_2 and increase the nitrogen content of the soil. During their growth most blue-greens release into their environment substantial quantities of extracellular products, mostly polysaccharides and peptides. They further enrich the soil in organic matter on their death and subsequent decomposition. The released substances support the activities of the bacterial and fungal flora and promote the mineralization of organic matter. The processes also contribute to the improvement of the physical-chemical properties of the soils (see also § 7.5).

4.7 Gliding movement

Many blue-greens, when viewed under the light microscope, show a variety of movements, such as gliding, rotation, oscillation, jerking and flicking. All such movements occur when trichomes, hormogonia or cells are in contact with a solid substratum. They appear to represent variations of locomotion which operate on the same principle, known as gliding movement.

Gliding is the smooth snail-like progression of filamentous blue-greens which proceeds with no visible signs of contraction or propulsion. It is the characteristic motion of epiphytic, benthic and terrestrial species which usually thrive within dense mats attached to some solid substratum, like rocks, submerged plants or waterlogged soil. Gliding is a relatively slow movement (its speed varies between about 1 and 10 μm s^{-1}), and because of this inherent sluggishness, gliding is scarcely important as a means of dispersing the species over a large area. It may, however, help the organism to adjust its position in the environment, by gliding forward and backward, and to settle in an area where conditions are favourable for growth.

Gliding motility is influenced by a number of environmental factors, like temperature, photon flux density and quality of light, pH and the chemical composition of the medium. The temperature dependence of gliding suggests the involvement of enzymatic reactions. Photosynthetic and respiratory metabolisms coupled with ATP synthesis are apparently the means whereby energy is provided for gliding motility. Certain species glide at the same rate in light and dark but gliding in the dark will depend on carbon reserves produced in prior photosynthesis and on the organisms ability for heterotrophic nutrition. Light could also determine the direction of movement. Filaments may glide towards the light source (positive phototaxis) but strong light may bring about a phobic response (photo-phobotaxis) and cause the filaments to stop and subsequently reverse. The photo-phobotactic response is caused by a sudden change (step-up or step-down) in the quantum dose of light. If a light field is projected through a slit onto a preparation of *Phormidium* on an agar plate (Fig. 4–7), the gliding filaments which eventually enter the light field from the surrounding dark field are unable to leave it because they respond by a photophobic reversal of movement each time when they pass the light-dark border. As a result, the filaments gather in the illuminated area which functions as a light trap. The response is termed 'step-down' or positive photo-phobotaxis. Increasing the quantum dose of light, the filaments may respond in a different way: they may leave the field and congregate in a fringe around the light field; such response is called 'step-up' or negative photo-phobotaxis. Action spetra of photophobic responses suggest that pigments of both Photosystem I and II participate in the absorption of light which stimulate phobic reactions. Photoreceptor and motor are probably located in different parts of the cell, hence the primary photoprocess must produce a transportable signal to bridge the gap. This has been termed sensory transduction. Signal transduction must occur not only within the cell but also between cells of the trichome. Sensory transduction is probably brought about by changes in the electric potential of the cytoplasmic membrane.

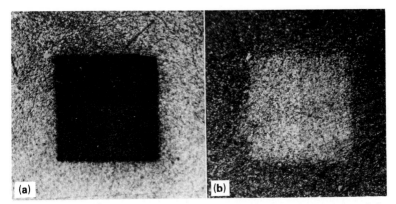

Fig. 4–7 (a) Step-down (positive) and (b) step-up (negative) photo-phobotaxis by *Phormidium*. Courtesy of Dr D.P. Häder.

In the Oscillatoriaceae gliding is accompanied by a right or left handed rotation of the trichome around its axis. The well-known swaying movement of *Oscillatoria* species is a type of gliding by a trichome with slightly bent non-adhering tips. The tips move in a spiral path as the trichome rotates, and give the false impression of swaying or 'oscillation'. Trichomes and hormogonia in Nostocaceae glide without rotation.

Filament aggregation or clumping arises from the gliding of trichomes over each other and through clusters of trichomes in opposite direction, thereby drawing together the filament clusters. A series of similar actions result in a mass contraction and aggregation of trichomes at a speed which can exceed many times the basic gliding rate. Clumping in nature enables mats to contract rapidly and thus to avoid excessive illumination by means of self-shading, and in thermal streams to withdraw from waters of adversely high temperature.

Gliding is often accompanied by mucilage production, and early interpretations of the gliding movement were based on the assumption that the directional secretion of mucilage may result in a slow displacement of the trichome. It was shown, however, that mucilage secretion is the consequence and not the cause of gliding movement. Nevertheless, mucilage has a useful function in providing adhesion between trichome and substratum.

More recent concepts relate gliding with the possible presence of contractile organelles, and are based on the discovery of numerous fine (5 to 8 nm wide) fibrous elements in the wall of the trichome. Circumstantial evidence suggests that these microfibrils are contractile and that synchronous waves propagated by the contractile fibrils propel the filament. The possible involvement of contractile elements is further indicated by the finding that gliding is effected (in *Oscillatoria terebriformis*) by calcium concentration. (Calcium is known to play part in muscle fibre action and in nerve cell function.) Also, jerking and clumping (in *Spirulina*) is markedly stimulated by the addition of neurotransmitters (like acetylcholine or noradrenaline), known to enhance the contraction of smooth muscle fibres.

5 Blue-greens in Symbiotic Associations

Blue-greens engage in symbiotic associations with a wide range of organisms. They are present in many lichens, in liverworts, in the aquatic fern *Azolla*, in the roots of cycads, in the glands of the angiosperm *Gunnera* as well as in a number of invertebrates. In these, blue-greens constitute the micro- or endo-symbiont, enclosed within specialized organs, tissues or cells of the macrosymbiont or host. Blue-greens also form casual loose associations with a variety of organisms, ranging from bacteria to higher plants.

All biological associations involve in interactions between the organisms concerned, in the interchange of metabolites, and result in a certain degree of interdependence. Blue-greens in their associations with other organisms represent in nutritional terms the productive partner. With non-photosynthetic organisms, they usually function as primary producers of organic matter. In a great number of associations the contribution of blue-greens is primarily through their ability to fix N_2. This appears to be their main function in symbioses with other photosynthesizing green plants.

5.1 Casual associations

A great proportion of bacteria associated with phytoplankton live attached to the algae and only a few are truly planktonic. Many bacteria are embedded in the often extensive mucilagenous sheaths surrounding blue-greens. These bacteria are adapted to the micro-environment of the mucilage envelope, and thrive on the extracellular organic products released and on the O_2 liberated by the algae. Bacterial assimilation of organic substrates results in the production of CO_2 which could be immediately available to the algae and re-assimilated in photosynthesis.

Many blue-greens are engaged in a reverse relationship by being attached to freshwater plants, sea grasses or seaweeds. Some of the epiphytic blue-greens have been shown to fix N_2. Transfer of fixed carbon from host to epiphyte and of fixed nitrogen in the reverse direction is a possibility but it has not yet been demonstrated. Unicellular and filamentous blue-greens (of the genera *Synechocystis* and *Phormidium*) frequently form bright patches over the surface of calcified ascidian (sea squirt) colonies. Little is known of these associations which seem to be highly specific.

5.2 Association between fungi and blue-greens

An interesting but little studied association develops between the phycomycetes fungus *Geosiphon pyriforme* and *Nostoc sphaericum* in the form of minute (about 0.2–0.3 mm in diameter) shiny dark vesicles. Both partners can grow independently, and the association is re-established afresh upon contact between the fungal hyphae and the *Nostoc* trichome; the hypha responds by the invagination of its plasma membrane and gradually encircles the trichome. Growth continues within the association and an equilibrium is maintained.

The lichen thallus is a uniquely integrated symbiotic system in which interaction between the fungal partner or mycobiont and the algal partner or phycobiont results in the formation of a distinct morphological entity. Most lichens contain only one kind of phycobiont which belongs either to green algae or to the blue-greens but in some lichens the thallus incorporates two different phycobionts. The secondary phycobiont is invariably a blue-green, and is found either in separate branches of the thallus or more often it occurs in discrete external or internal protuberances called cephalodia (Fig. 5–1). *Nostoc* is the most common blue-green in lichens but the genera *Calothrix*, *Dichothrix*, *Stigonema* and *Scytonema* are also represented.

Much useful information of the relationship between the symbiotic partners has been inferred by comparing the physiology of the intact thallus with that of the isolated components. Rates of photosynthesis and N_2-fixation were found to be several times higher in the intact thallus of the lichen *Peltigera* compared to the rates observed with the isolated *Nostoc*. Short term isotope tracer experiments with intact *Peltigera* have confirmed rapid transfer of freshly fixed carbon from alga to fungus. Within two minutes of being fixed, carbon is released in the form of glucose, assimilated by the fungus and converted to mannitol. A large proportion of N_2 fixed by *Nostoc* in cephalodia is translocated to the fungal hyphae in the form of ammonia (Fig. 5–2). Only a small fraction of the fixed nitrogen is retained by *Nostoc* filaments, and none is received by the green phycobiont, *Coccomyxa*. It is thought that the fungus may be able to modify the permeability properties of the *Nostoc* plasma membrane by chemical action. Increased permeability would enhance the movement of metabolites, sugars and ammonia, from the algal cells to the fungus.

N_2-fixing activity of the blue-green partner in polysymbiotic lichens is greatly affected by the presence and proximity of the green phycobiont. In lichens with a blue-green phycobiont only, the relative number of heterocysts (about 5%) is similar to that observed in free-living forms. In lichens incorporating both a blue-green and a green alga, however, heterocyst frequency in the blue-green filaments may be 5 to 10 times higher and the rates of N_2-fixation subantially elevated. Ultrastructural and biochemical evidence indicates that *Nostoc* filaments in *Peltigera* are kept in a nitrogen-starved condition probably as a result of continuous leakage from their cells of nitrogenous products. Prolonged nitrogen depletion enhances heterocyst differentiation and the synthesis of nitrogenase. Furthermore, a direct relationship has been established between N_2-fixing activity (heterocyst frequency) of *Nostoc* and the proximity of *Coccomyxa*. This could be the result of an augmented supply of photosynthetic products from *Coccomyxa* to

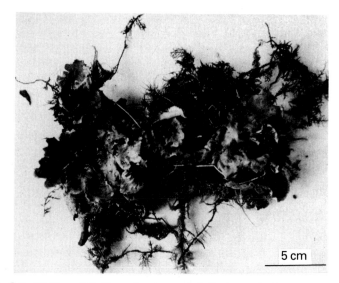

Fig. 5–1 Thallus of *Peltigera* showing cephalodia (arrows). Courtesy of Dr J.W. Millbank.

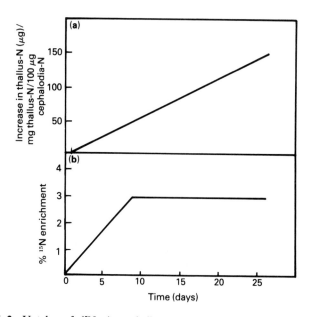

Fig. 5–2 Uptake of $^{15}N_2$ into thallus and cephalodia of *Peltigera aphthosa*. (a) Nitrogen uptake by the thallus is linear. (b) ^{15}N enrichment of the cephalodia exposed to $^{15}N_2$ became constant after 10 days (after Millbank and Kershaw, 1969, *New Phytol.*, **68**, 721–9).

Nostoc. Blue-greens in symbiotic associations are distinguished by their heterotrophic abilities. Sugars released by the green alga may be taken up not only by the fungus but also by *Nostoc* if located near *Coccomyxa*. Assimilation of organic compounds by *Nostoc* will in turn enhance heterocyst development and nitrogenase synthesis.

Physiological conditions in the lichen thallus so far described are clearly beneficial to the fungus and may appear rather disadvantageous to the blue-green partner. However, this is counterbalanced by certain other factors like water supply, mineral nutrition, optimal light conditions and reduced O_2 tensions essential to the phycobiont. It has been shown that O_2 tension within the lichen thallus is generally below atmospheric levels. Such conditions, probably maintained by active fungal respiration, may considerably promote N_2-fixation in *Nostoc*.

5.3 Symbioses between liverworts and blue-greens

Blue-greens form symbiotic associations not only with heterotrophs but also and rather unexpectedly with photosynthetic organisms. The associatons between *Nostoc* and the bryophytes *Blasia*, *Anthoceros* and *Cavicularia* have been known for a long time and have been investigated in some detail. The blue-green endosymbiont occupies mucilage-filled intercellular cavities within the liverwort thalli. The invading *Nostoc* filaments enter through small mucilagenous pores.

A colony of closely-packed *Nostoc* filaments develops within a few days, and this induces the host cells surrounding the cavity to form thread-like protrusions into the *Nostoc* colony (Fig. 5–3). The protuberances greatly increase the bryophyte-*Nostoc* interface promoting the interchange of metabolites.

Once installed inside the liverwort cavity, the *Nostoc* colony becomes metabolically specialized for the function of N_2-fixation and depends almost completely on the host for a source of carbon and energy. Fixed carbon, probably as sucrose, is liberated by the host cells and translocated to the endosymbiont. The *Nostoc* cells are depleted of nitrogenous storage products, cyanophycin granules and phycobiliproteins. The relative number of heterocysts rises to 30% and often to 50%, and the endosymbiont is actively engaged in N_2-fixation (Table 5). However, it fixes very little CO_2 and liberates no O_2.

It is not difficult to separate the individual partners and to grow both *Nostoc* and the liverwort in isolation. *Nostoc* colonies when released from *Anthoceros* or *Blasia* liberate substantial amounts of ammonia and show little if any ability to fix CO_2 during the initial hours of isolation, but can readily assimilate sugars and may even grow heterotrophically in the dark.

5.4 Symbiosis between a fern and a blue-green

The association between the floating water fern *Azolla* and its *Anabaena* endosymbiont is the only known example of a symbiotic relationship between a

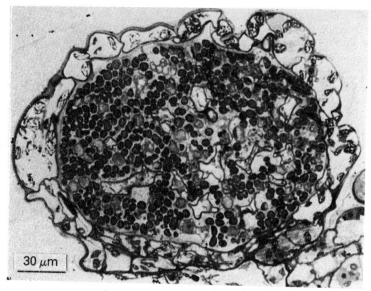

Fig. 5–3 Section through a domatium of *Blasia* with *Nostoc* filaments. Courtesy of Prof. J.G. Duckett.

fern and a blue-green (Fig. 5–4). The association has many of the features which characterize the *Nostoc*-liverwort symbioses. Here also both components are photosynthetic organisms, and the blue-green is adapted to the principal function of N_2-fixation. It penetrates the fern leaf through pores on the ventral surface and inhabits cavities in the dorsal leaf lobes of the fern (Fig. 5–4b). Each cavity is lined with a thin envelope from which multicellular hairs protrude into the cavity space. The hairs may fulfil a similar function in the exchange of metabolites between fern and *Anabaena* as the protuberances in the cavities of *Anthoceros*. The growth of *Anabaena* in the cavities proceeds with *Azolla* leaf tissue development. In young leaf cavities the blue-green cells

Table 5 The influence of symbiotic association with *Blasia pusilla* on heterocyst formation and nitrogenase activity (acetylene reduction) of *Nostoc* (after Rodgers and Stewart, 1977, *New Phytol.*, **78**, 441–58).

Condition of Nostoc	Heterocyst frequency (%)	Nitrogenase activity (n moles C_2H_2 reduced mg protein^{-1} h^{-1})
Free-living isolate	6	22
2 weeks ⎱ following re-association	20	54
6 weeks ⎰ with *Blasia*	48	70

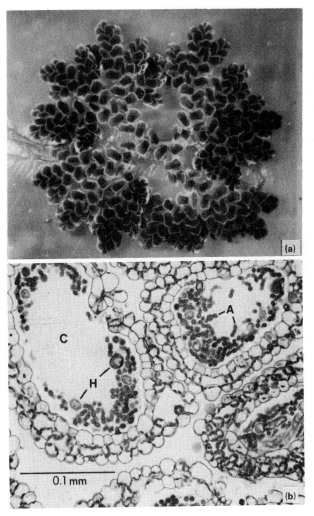

Fig. 5–4 (a) *Azolla* plant (b) Section through a leaf cavity (C) of Azolla containing *Anabaena* filaments (A) and epidermal hairs (H). Courtesy of Dr G.A. Peters.

are small and rapidly dividing, the filaments lack heterocysts and display no nitrogenase activity. As the leaf grows the cavities become colonized with *Anabaena* filaments which begin to develop heterocyst and fix N_2. At a stage when the leaf is about 12 nodes (leaf distances) from the thallus apex, heterocyst frequency rises to 30–40%. Both Photosystem I and II of the *Anabaena* are suppressed in the symbiotic condition. When isolated, however, *Anabaena* is capable of photosynthesis and performs light-dependent N_2-fixation though at a decreased rate.

The *Azolla–Anabaena* association is one of the most efficient N_2-fixing systems. The exceptionally high rates of N_2-fixation measured are almost certainly attributable to the abundant supply of photosynthetic products from the host to the N_2-fixing endosymbiont, which permits N_2-fixation to continue at night. The beneficial application of *Azolla* as a nitrogen fertilizer in rice fields of South-East Asia has been known for some time. Modern systems of cultivation and application of *Azolla* are discussed in section 7.5.

5.5 Blue-greens associated with higher plants

Two groups of seed-bearing plants, the gymnosperm group of cycads and the angiosperm genus *Gunnera*, harbour heterocystous blue-greens.

In the root nodules (or 'coralloid' roots) of cycads, nostocacean filaments occupy the mucilage-filled intercellular spaces in the distinctly green middle zone of the root cortex. Following invasion, the cells bordering this intercellular space produce tubular outgrowths and secrete at least part of the mucilagenous matrix. Although the formation of coralloid roots is independent of the presence of *Nostoc*, only those which are invaded by the blue-green persist. Both the host and the endosymbiont can grow in isolation.

In *Gunnera* the endosymbiont *Nostoc punctiforme* is located in wart-like swellings or nodules of the stem near the leaf bases. Infection is initiated by the host papillate gland which secretes mucilage and this greatly stimulates the growth of *Nostoc*, which penetrates the thin-walled meristematic cells at the base of the gland in the form of gliding hormogonia and rapidly develops within the nodule. Thus the blue-green filaments occur within the host cell enveloped in the host plasma membrane. A large proportion of cells (up to 80%) transforms to heterocysts. The endosymbiont fixes N_2 at a rate which meets the total nitrogen requirement of the growing host plant. Again, the host and the endosymbiont can be grown separately. The symbiosis has been reconstructed successfully in the laboratory.

5.6 Symbioses of blue-greens with invertebrates

Blue-greens are known to occur in an echiuroid worm and in several marine sponges. Their presence in the latter is recognizable from the green colour of the sponge cortical tissue (ectosome). Aggregates of a unicellular blue-green (possibly *Aphanocapsa*) occur in specialized vacuolated cells of the sponge, called cyanocytes (Fig. 5–5). There is good indication that the endosymbiont is able to fix N_2 within the sponge system. This may be particularly beneficial to sponges in tropical seas which are generally short of combined nitrogen. Examination of the sponge tissue in the electron microscope shows very few dividing blue-green cells, and this may suggest that endosymbiont reproduction and metabolism is under host control, effecting the translocation of nutrients from endosymbiont to sponge. It seems remarkable that most blue-green cells remain intact. Sponges are known to be omnivorous filter feeders and one would expect blue-greens to be liable to digestion within the sponge.

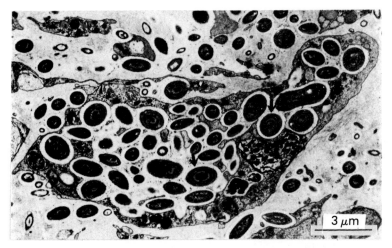

Fig. 5–5 Endosymbiotic blue-greens cells (arrows) in a mesophyll cell of the sponge *Siphonochalina*. Courtesy of Dr C.R. Wilkinson.

5.7 Cyanelles and the origin of chloroplasts

Several eukaryotic protists contain green chloroplast-like structures in their cytoplasm which resemble unicellular blue-greens. They were called cyanelles and are generally considered to represent endosymbiont blue-greens which have assumed the function of a photosynthetic organelle within the host cell. *Cyanophora paradoxa*, a biflagellate protist contains 2 to 4 spherical cyanelles which were named *Cyanocyta korschikoffiana* (Fig. 5–6). The rod-shaped cyanelle in *Glaucocystis nostochinearum* was called *Skujapelta nuda*. The justification for this taxonomic distinction of chloroplast-like structures, however, is doubtful.

The cyanelles of *Cyanophora paradoxa* possess a photosynthetic system which in its structure and composition resembles that of blue-greens. Movement of photosynthetic products from the cyanelle to the host cytoplasm has been demonstrated in experiments using radioactive carbon. Electron micrographs show that the cyanelles are surrounded by a single membrane. The presence of an external thin peptidoglucan envelope, however, which is characteristic of blue-greens, has been substantiated by the finding that isolated cyanelles retain their shape and are extremely sensitive to treatment with the enzyme lysozyme, known to degrade peptidoglucan. Cyanelles divide within the host and are evenly distributed between the daughter host cells but their division is not in synchrony with that of the host. The genetic complement of the cyanelle resembles that of the chloroplast genome in size and in DNA-base composition, more than that of a blue-green cell. The association between host and cyanelle appears to be mutually obligate as

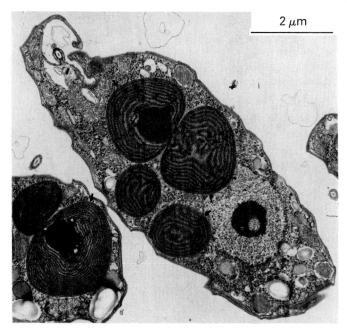

2 μm

Fig. 5–6 Cyanelles (arrows) in *Cyanophora*. Courtesy of Prof. R.K. Trench.

neither of the partners can be grown in isolation. The cyanelle is therefore regarded a photosynthetic organelle analogous to the chloroplast.

The existence of cyanelles is consistent with the endosymbiont hypothesis of the evolution of chloroplasts. This hypothesis assumes that chloroplasts were derived from photosynthetic prokaryotes like blue-greens which adopted an endosymbiotic existence within a phylogenetically unrelated heterotrophic ancestral eukaryotic cell. Cyanelles then would represent intermediate forms and could resemble more closely the ancestral photosynthetic prokaryotes. The hypothesis is supported by the structural similarities between chloroplasts and blue-greens, by the similarities between chloroplast and blue-green cell ribosomes and ribosomal RNAs, and by the almost identical composition and function of the photosynthetic apparatus in chloroplasts and blue-greens.

6 Genetics, Evolution and Taxonomy

Blue-greens display a great diversity of forms which differ in their morphology, structure and function, and in their mode of response to environmental stimuli. Their characteristics are expressions of their genetic make-up (genotype) which itself is the result of many million years of evolutionary history. It is important to explore the genetic constitution of an organism and to establish how this is expressed in the structure and function of the organism, how this expression is regulated, and how it is effected by environmental conditions. A comparative study of the genetic constitution of blue-greens and the assessment of genetic relationships between blue-greens and other groups of micro-organisms is the basis of a comprehensive revision of the taxonomy of blue-greens.

6.1 The occurrence of mutations and the utility of mutants

Mutations, permanent changes in the quality (chemistry) or quantity of the genome, may occur 'spontaneously' but recognition and selection of mutants in blue-greens is often difficult. Nevertheless, records of the occurrence of spontaneous mutants in laboratory cultures are well substantiated. These include strains which fail to differentiate heterocysts and to synthesize nitrogenase (*Anabaena variabilis*), which lack the ability to produce akinetes (*Anabaena cylindrica*), which are incapable of forming gas vesicles (*Anabaena flos-aquae*), of synthesizing toxins (*Microcystis aeruginosa*) or of depositing sheath material. Such mutants generally outgrow the wild type because energy previously invested in the formation of structures which confer no advantage in particular conditions can be diverted to cell growth. Stable drug and antibiotic resistant mutant stains, which display hundred to several thousand times higher resistance, to the chemicals than the wild type, have been selected by successive transfers into media supplemented with gradually increased concentrations of the chemical. The drug resistance of mutants has been used as a marker in studies on the occurrence of genetic recombination in blue-greens. Mutations can be induced artificially at a much higher frequency by the use of chemical agents or by exposure to physical agents (ultraviolet light or ionizing radiation) which produce chromosome aberrations.

A variety of morphological mutants has been isolated. Filamentous mutants of unicellular form (*Synechococcus*, *Agmenellum*), possessing or lacking cross walls, have been used in studies of genetic regulation of cell division and septum formation. Mutants displaying altered or irregular patterns of heterocyst spacing have been employed in studies of the control of heterocyst differentiation and akinete formation. Ultrastructural mutants, like those defective in the synthesis of the normal heterocyst envelope and containing a highly O_2-sensitive nitrogenase, are useful in assessing the function of a particular cellular structure. The occurrence of mutants producing heterocysts but failing to synthesize nitrogenase suggests that the two processes are partially independent; they support the hypothesis that the heterocyst is a more recent evolutionary acquisition by blue-greens which serves to protect the more ancient nitrogenase system against increasing hazards of inactivation by atmospheric O_2.

Different classes of biochemical mutants have been produced and characterized. They include metabolic mutants which are deficient in solute transport, which lack some of the key enzymes involved in the assimilation and metabolism of essential nutrients, or which fail to produce enzymes required for the endogenous breakdown of macromolecules. Mutants deficient in an active nitrate reductase or nitrogenase have been used to establish the pathways of nitrogen assimilation in blue-greens. Pigment mutants are useful in the examination of the role and function of individual pigments of blue-greens. The mechanism of action of inhibitors can be studied with mutants resistant to the action of such compounds. For example, two of *Aphanocapsa* strains, both resistant to the action of DCMU, were found to differ in the nature of their resistance; in one strain this was attributable to a decreased permeability to the herbicide, in the other strain to alteration in the photosynthetic electron transport system, probably near the site of the inhibitory action of DCMU. Auxotrophic mutants (which grow only in the presence of a particular organic compound they are unable to synthesize) have been used to determine biosynthetic pathways and in studies on the regulation of gene expression in blue-greens.

6.2 Gene transfer and the presence of plasmids

During sexual reproduction of eukaryotic organisms the parent gametes each contribute a single set of unpaired chromosomes to the chromosome complement of the diploid cell and to the genome of the progeny. No similar process of sexual reproduction is known to occur in the prokaryotes. However, in the course of 'parasexual' phenomena partial genetic transfer to a recipient cell is possible, although this process is independent of reproduction.

Of the three known modes of genetic transfer in bacteria (conjugation, transformation and transduction), only transformation (genetic recombination by means of DNA released, or extracted, from a donor cell entering into a recipient cell and replacing part of the recipient genome) has been shown conclusively to occur in a number of blue-green species. Cells of an erythromycin-sensitive strain of *Anacystis nidulans* became erythromycin-resistant by apparent transfer of DNA extracted from cells of the antibiotic-resistant strain. The occurrence of intergeneric gene transfer (between *Gloeocapsa alpicola* and *Anacystis nidulans*) has also been reported.

Many blue-greens have plasmids (extrachromosomal covalently-closed circular DNA molecules) which are fragments of DNA of viral or bacterial origin that lie in the host cytoplasm and replicate independently. Plasmids appear to be quite common companions of the host blue-green genome: more than half of the strains tested contained one or more plasmids. Interestingly, plasmids of the same characteristics are present in different strains of blue-greens; this seems to indicate that plasmid transmission has taken place between strains of the same genus as well as between different genera. Exchange of genetic information by viral or bacterial transduction, however, has not yet been demonstrated in blue-greens.

The role of plasmids in the phenotypic expression of inherited properties, in the growth and natural distribution of species, or in the evolution of blue-greens is as yet unknown. There are, however, certain indications of plasmid-determined characteristics being expressed in the natural phenotype, particularly those which may be readily lost in culture, such as, for example, the ability to produce toxins or to form gas vesicles, or the resistance to heavy-metal poisoning. Plasmids might also be used as carriers in gene transfer systems and may aid the genetic analysis of blue-greens.

6.3 The fossil evidence of cyanobacterial evolution

Palaeobiology has recently provided the long sought for and most convincing records of remarkably well preserved microfossils that appear to represent the earliest known prokaryotic micro-organisms. They were discovered in fine grained chert deposits and in anceint stromatolites (characteristic sedimentary structures produced in near-shore environment and composed of alternating calcareous and carbonaceous layers incorporating blue-greens and flexibacteria), which were formed in the Precambrian period between about 3.2 and 0.6 billion years ago. As fossil unicellular and filamentous blue-greens predominate in these sediments, the Precambrian era has been named 'the age of blue-green algae'.

The oldest sediments from the Early Precambrian rocks, about 3.1 billion years old, contain mineralized structures which resemble present-day unicellular blue-greens. The microbial population gradually diversified during the period and began to appear in growth forms characteristic of modern unicellular genera. Simple filamentous forms of the oscillatorian type were also discovered in the Early Precambrian stromatolites about 2.8 billion years old. The Middle Precambrian assemblage of microfossils is dominated by heterocystous forms, mainly of the nostocacean and rivularian type. Their occurrence in the Middle Precambrian sediments constitutes palaeobiological evidence for the transition at this period (between 2.5 and 1.7 billion years ago) of the Earth's atmosphere from relatively O_2-free to a progressively oxygenic nature, apparently as a direct result of cyanobacterial photosynthesis (seé § 3.7).

Palaeobiological and geochemical records indicate that blue-greens have reached their peak in the Mid-Precambrian era about 2 billion years ago. Their dominance began to decline with the rise of eukaryotic algae about 1.6 billion years ago. The development of aerobic metabolism and of eukaryotic sexual reproduction greatly enhanced the speed of biological evolution. Regular genetic recombination resulted in greatly increased species variability and brought about the rapid diversification of the evolutionary tree.

6.4 Genealogy and prokaryotic taxonomy

Genealogy, the study of biological evolution, has progressed through the concordant and contrasting evidence provided by palaeobiology and comparative analytical biology. The development of phylogenetic schemes has in turn influenced contemporary taxonomy. Modern biological classification is based on natural relationships between different taxa and the various groups of organisms. Classical taxonomy made use of easily recognizable, though not necessarily representative, morphological characters for the identification and classification of plant and animal forms. This approach has on the whole failed with bacteria, where shape and form are much less distinct and expressive characteristics. Instead, various cytochemical and biochemical tests were devised for the identification of bacteria, which unfortunately were seldom suitable for establishing the true relationships between different strains.

The taxonomy of blue-greens rests mainly on the description and identification of morphological characteristics, which in general are far more distinct and recognizable than those of bacteria. Characterizations were routinely made in the past on specimens collected in the field. This practice, however, introduced a measure of uncertainty as to the true value of morphological characteristics as taxonomic tools, since natural material is prone to considerable morphological variability depending on the prevailing environmental conditions. On the other hand, characterization based exclusively on axenic cultures may be equally inadequate, as the limited and constant physical-chemical conditions applied for the growth of laboratory cultures may restrict the full expression of the genome. A critical comparison between material collected from the natural habitat and that grown in isolation and under controlled laboratory conditions is clearly called for. It is also important to take into consideration characteristics other than morphology. Indeed, the use of as many well definable (structural, biochemical, physiological and ecological) characteristics as possible should aid to establish the whole spectrum of phenotypical expressions of a particular genome. These commendable objectives, however, demand a great deal of further analytical work and computing before the system may become suitable for field application.

6.5 Macromolecular approach to prokaryotic phylogeny

Biological macromolecules, primarily proteins and nucleic acids, are known to exhibit common group characteristics as well as a high degree of specifity. Genetic information for the entire complement of phenotypically-expressed characteristics is incorporated in the nucleotide sequence of DNA, and thus the relatedness between DNAs of different organisms should reflect their genealogical relationships. This was neatly expressed in a statement that 'macromolecular sequences betray evolutionary history', and similar considerations are the reason for the growing interest in the assessment of genetic relatedness of micro-organisms by means of comparative analysis of macromolecules. Comparative analysis of amino acid sequences of key proteins has been used successfully to establish the pathways of evolution in the eukaryotes. Phycobiliproteins would be the obvious candidates for comparative amino acid sequencing to establish phylogenetic relationships among blue-greens.

There are several ways of using DNA for the assessment of genetic relatedness of blue-greens. One is the determination of genome size (see § 1.3). It has been shown that genome size varies considerably among the blue-green strains tested. Nevertheless, the finding that most unicellular strains possess relatively small genomes (similar in size to those of bacteria), and that the pleurocapsalean and filamentous forms have larger genomes (probably acquired by means of duplication of the ancestral genome), seems to support the conclusion drawn from palaeontological evidence that the unicellular forms were first to evolve and that they gave rise in Precambrian times to other forms with more complex morphology.

DNA base composition (see § 1.3) has been used as a genotypic character but the range of mean DNA base composition in blue-greens, particularly among unicellular forms, is almost as wide (between 35 and 71 mole % GC) as that found in all other prokaryotes. DNA base composition determinations thus seem to have a limited value in the assessment of generic and specific interrelations and delimitations.

A comparative study of nucleotide sequences of DNA appears to be more promising but is only possible by means of an indirect approach, DNA–DNA hybridization, due to the lack of suitable techniques for the sequential degradation of DNA. The hybridization technique is based on the observation that single strands of DNA separated by heating will re-associate after cooling and rebuild the original double helix. Re-association is 100% when the two DNA strands were derived from one and the same organism or strain, whereas strands from different strains associate (hybridize) only in regions of homologous nucleotide sequences. The extent of hybridization can be assessed by means of previous radioactive labelling of one of the single DNA strands, so that the amount of radioactivity present in the hybrid DNA is an indication of the degree of homology. A similar technique can be used to determine to what extent RNA, derived from one organism, is complementary to the single strand of DNA, extracted from another organism. These techniques have provided some interesting information concerning genetic relatedness between blue-greens, bacteria and chloroplasts. Very low degrees of cross DNA hybridization have been observed between bacteria and blue-greens but significant degrees of hybridization (up to 47%) have been obtained between chloroplast DNA and ribosomal RNAs from various blue-green strains, suggesting a close genetic relationship between chloroplasts and blue-greens (see § 5.7). The degree of hybridization between different heterocystous species is relatively high (up to

40%) while binding efficiencies between heterocystous and non-heterocystous forms varied between 10 and 20%.

It appears that by far the most expedient and reliable method of determining prokaryotic genealogies is by means of comparative analysis of oligonucleotide sequences released by T1 ribonuclease digestion of 16S ribosomal RNAs (the rRNA species of the smaller component of prokaryotic ribosomes having a sedimentation coefficient of 16 Svedberg). The enzyme (extracted from the fungus *Aspergillus oryzae*) catalyses the cleavage of rRNA to yield oligonucleotides, which are then separated and analysed to determine their nucleotide sequences. The pattern or 'catalogue' of oligonucleotide sequences of 16S rRNA is characteristic of a particular species. The comparison of catalogues reveals the phylogenetic relatedness between species tested, and this can be expressed quantitatively in terms of a similarity coefficient, S_{AB}. There are several reasons why rRNAs are superior to any other molecules for the assessment of phylogenetic relatedness: rRNAs are of universal distribution, they have apparently retained their function throughout phylogeny, and appear to change in their nucleotide sequences much more slowly than other macromolecules; they are not transmitted to other organisms, and can be easily isolated. The method has been applied to a great number and variety of bacterial and blue-green species, and has yielded much important information regarding the evolution of the Prokaryota as well as to the evolutionary pathways within the blue-greens, which will certainly have a profound influence upon prokaryotic taxonomy.

6.6 The macromolecular phylogeny of blue-greens

Comparative analysis of 16S ribosomal RNA sequences has been successfully applied to reveal the phylogenetic affinities among blue-greens and other major groups of prokaryotes, as well as between blue-greens and chloroplasts. On the basis of homologies (Fig. 6–1 a and b) several important postulates can be made:

 (*i*)   Blue-greens are phylogenetically remote from the best studied bacteria, though nearer to the bacilli than to the enteric bacteria.

 (*ii*) Chloroplasts of red algae (*Porphyridium*) are closely related to blue-greens but those of *Euglena* and of high plants (*Lemna*) are more distant and possibly not of direct blue-green origin.

 (*iii*) The unicellular genera *Synechococcus*, *Aphanocapsa* and *Agmenellum* represent a diverse but distinct group of most ancient blue-greens.

 (*iv*) Ancient unicellular blue-greens similar to present day *Synechococcus* strains have given rise to heterocystous forms.

 (*v*)   The Stigonematales are closely related to Nostocaceae in spite of their more complex morphology.

According to a general scheme of prokaryotic phylogeny (Fig. 6–2), the common ancestor of all extant organisms has given rise to three principal lines of development. The first lineage leads through the ancestral archaebacteria to

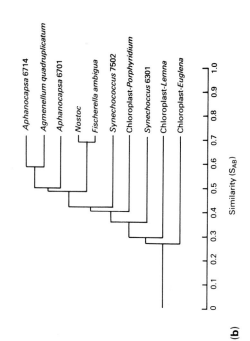

Fig. 6–1 (a) The phylogenetic relationships among the true bacteria (eubacteria), cyanobacteria and chloroplasts based on comparative analysis of 16S ribosomal RNA sequences. (b) Phylogenetic relationships among the blue-greens and chloroplasts (after Fox *et al.*, 1980, *Science.*, **209**, 457–63).

(a)

Similarity (S_{AB})

Bacillus subtilis
Lactobacillus brevis
Clostridium pasteurianum
Mycoplasma capricolum
Arthrobacter globiformis
Actinomyces bovis
Streptomyces griseus
Rhodospirillum tenue
Rhodopseudomonas sphaeroides
Rhodomicrobium vannielii
Escherichia coli
Chromatium vinosum
Desulfovibrio desulfuricans
Spirochaeta aurantia
Spirochaeta halophila
Aphanocapsa 6714
Synechococcus 6301
Chloroplast (Euglena)
Micrococcus radiodurans
Chlorobium limicola
Chloroflexus aurantiacus

(b)

Similarity (S_{AB})

Aphanocapsa 6714
Agmenellum quadruplicatum
Aphanocapsa 6701
Nostoc
Fischerella ambigua
Synechococcus 7502
Chloroplast-Porphyridium
Synechococcus 6301
Chloroplast-Lemna
Chloroplast-Euglena

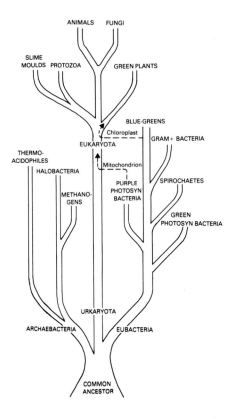

Fig. 6–2 Scheme of evolutionary pathways among the prokaryotes and the origin of eukaryotic organisms (after Woese, 1980, *Sci. Amer.*, **244**, 94–106).

methano-, halophilic and thermoacidophilic bacteria. The second major evolutionary line is represented by the true bacteria (eubacteria), diversifying into the various groups of photosynthetic and non-photosynthetic bacteria, and includes the cyanobacteria. The third line of evolution proceeded through a hypothetical ancestral 'urkaryote' group (having 18S ribosomal RNA) to eukaryotic organisms. The true eukaryotic cell is seen as a phylogenetic chimera, which incorporates eubacterial components (mitochondrion, chloroplasts) of endosymbiotic origin in the urkaryotic cytoplasm.

7 Economic Importance

The abundant growth of blue-greens may be a source of considerable nuisance to man in some situations, as in the management of water reservoirs, while it is of great importance in other areas, like in the maintenance of soil fertility in rice fields. This chapter examines both the detrimental and beneficial activities of blue-greens. It reviews some of the practical means available to control their growth where they are undesirable. It also considers the prospects of their exploitation for food and energy production.

7.1 Algal nuisance in water supplies

Extensive growth of algae can create severe problems in the maintenance of water supplies and in meeting the ever increasing demand for potable water. Water bodies in mountainous regions are usually poor in nutrients (oligotrophic) and can support the growth of only a small algal population. But water in lowland storage reservoirs, often fed with water from rivers and streams and frequently very rich in nutrients (especially phosphorus), may yield exuberant algal growth and typical blooms of blue-greens.

The problems caused by dense algal populations are many fold. Firstly, they impede the routine filtration procedure essential in the provision of potable water. Large algal blooms may rapidly clog not only the fine sand filters but even the primary course fast filters. Secondly, algae may release substances in the water which are harmful or toxic, which cause unnatural colouration of the water or which add an objectionable odour and/or taste to drinking water. (Toxins produced by blue-greens are discussed below in more detail.) Lysing cells of blue-greens release the water-soluble phycobiliprotein pigments, giving the water a bluish or pinkish colour. Planktonic blue-greens have a quite pleasant grassy odour while healthy and intact, but this may change to an unpleasant musty smell or to a rather revolting foul odour upon disintegration and bacterial decomposition of bloom-forming material. Furthermore, several blue-green species grow attached to reservoir walls and pipes restricting thereby water transport. Others may cause corrosion of concrete and steel structures. The mass appearance of blue-greens has also a distractive effect upon recreational activities, like swimming, sailing or fishing, in lakes and reservoirs.

7.2 Toxic blue-greens

Numerous episodes of poisoning of farm animals (such as cattle, sheep, horse and pig) and of waterfowl and wildlife were described throughout the world and attributed to the ingestion by drinking animals of lethal doses of toxic blue-greens accumulating at the surface of water or along the shores of temperate lakes, ponds and reservoirs. Several species have been indicated but toxic strains of three common bloom-forming species, *Anabaena flos-aquae, Microcystis aeruginosa* and *Aphanizomenon flos-aquae,* appear to be responsible for most, if not all, incidents of serious consequences.

An extremely powerful poison was isolated from *Anabaena flos-aquae* which kills mice upon intraperitoneal injection within 1–4 minutes; death is preceded by symptoms of paralysis, tremor and convulsion. It was first called 'very fast death factor' but later renamed anatoxin a. It is an alkaloid with a molecular weight of 165 daltons and a structure resembling that of cocaine (Fig. 7–1a). The poison acts as a potent neuromuscular blocking agent and causes death by respiratory arrest. The minimum lethal dose (LD_{min}) following intraperitoneal injection is 0.3 mg per kg body weight. It was estimated that this quantity of toxin can be delivered to cattle by the ingestion of about 3–4 litres of a dense suspension containing toxic *Anabaena flos-aquae*.

Two distinct toxins were isolated from *Microcystis aeruginosa*, one causing death within 48 hours (called 'slow death factor'), another killing mice within 1 to 3 hours (fast death factor). Further investigation has revealed that accompanying bacteria were responsible for the former toxic effect but *Microcystis* is the source of the 'fast death factor', now renamed microcystin. It is a low molecular weight (about 1500 daltons) cyclic polypeptide which yields 7 (or 14) amino acids upon hydrolysis. It causes enlargement and congestion of the liver followed by necrosis and haemorrhage, and may also exhibit neurotoxic activity. LD_{min} is 0.47 mg per kg body weight.

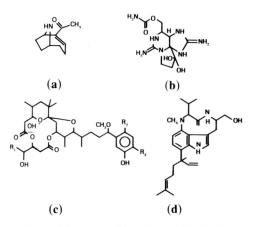

(a)   (b)

(c)   (d)

Fig. 7–1 The structures of (a) anatoxin, (b) saxitoxin, (c) lyngbyatoxin and (d) debromo-aplysiatoxin (after Gorham and Carmichael, 1979, *Pure and Appl Chem.,* **52,** 165–74).

The poison isolated from toxic blooms of *Aphanizomenon flos-aquae* was called aphantoxin. It appears to be a mixture of several toxic compounds, one being saxitoxin (a tetrahydropurine alkaloid; Fig. 7–1b) which is also responsible for paralytic shellfish poisoning. (The source of saxitoxin in this case is the flagellate *Gonyaulax catenella* which is ingested by the shellfish.) It is a fast-acting neuromuscular poison with LD_{min} of 10 mg per kg body weight.

Several freshwater and marine blue-greens were implicated in outbreaks of human gastroenteritis and in various forms of dermatitis. The frequently occurring 'swimmers itch' is attributed to contact with *Lyngbya majuscula, Schizothrix calcicola* and *Oscillatoria nigroviridis*, which are commonly found in tropical and subtropical seawaters. The toxins responsible are lipid-soluble phenolic compounds (Fig. 7–1c). The poisonous marine rabbitfish (*Siganus fuscescens*) and the sea-hare (*Stylocheilus longicauda*), a gastropod mollusc, both feed on *Lyngbya majuscula* and appear to accumulate toxic substances (aplysiatoxin and debromoaplysiatoxin), apparently derived from their blue-green diet (Fig. 7–1d).

Though most studies were directed to establish the effects of blue-green poisons on warm-blooded vertebrates and so far little is known of their influence on the natural grazers, there are good indications that blue-green toxins may act on zooplankton and might be an effective mechanism of protection against grazing pressures.

The reasons for the sporadic occurrence of toxic blooms are unknown, and this has given rise to a great deal of speculation. Toxic strains seem to be indistinguishable from non-toxic strains of the same species. It has been suggested that such strains may develop only under a particular set of environmental conditions, or that toxin production may be associated with plasmid-mediated gene transfer (see § 6.2).

7.3 Measures to control the growth of blue-greens

Chemicals are widely used to prevent the growth of nuisance algae, and the commonest treatment is the application of copper sulphate (at a concentration of $0.5–1.0$ mg l^{-1}). Although initially such low concentrations are not toxic to fish or other aquatic animals, repeated administration may increase the concentration of copper in the water to a toxic level. Planktonic blue-greens appear to be more sensitive to copper ions than diatoms or green algae, and hence lower concentrations of the salt are often adequate to suppress their growth. A number of other algicides are used effectively in water supplies, like phenolic compounds, amide derivatives, quaternary ammonium compounds and quinone derivatives. Dichloronaphthoquinone is selectively toxic to blue-greens. The hazards of using toxic chemicals indiscriminately in the natural environment are well documented, and hence it would seem advisable to replace such practices by other less hazardous measures and to use chemical control only as a last resort.

Biological control is in principle possible, though not always practical and as effective as chemical control. It can be achieved by the introduction of natural grazers and pathogens of blue-greens. Invertebrates like cladocerans, copepods, ostracods and snails are known to graze on green algae and diatoms. *Daphnia pulex* has been reported to feed on *Aphanizomenon flos-aquae* in

North American lakes, and it may prevent bloom formation while *Aphanizomenon* is present in the form of single filaments or small colonies but the cladoceran appears to avoid the large raft-like colonies. The cyclopoid copepod *Thermocyclops hyalinus*, common in Lake George, Uganda, feeds on blue-greens which dominate the lake phytoplankton, and it was clearly demonstrated that blue-greens are not only ingested but also digested and assimilated by the animals. The copepod *Diaptomus* has been implicated in the grazing of *Anabaena* populations in Severson Lake, Minnesota. Another copepod, *Cyclop vernalis*, feeds on *Oscillatoria* species dominating the shallow artificial lake in St. James Park, London. The ostracod *Cypris* would probably diminish the blue-green population of flooded rice fields if pesticides were not used to keep down its numbers. Among the vertebrates, the herbivorous fish *Tilapia nilotica* and *Haplochromis nigripinnis*, which are common in Lake George and represent the most important fish yields in the region, have been shown to feed mainly on blue-greens.

Micro-organisms (fungi, bacteria and viruses) appear to play an important part in regulating the growth of blue-green populations in freshwaters. In the English Lake District chytrids (fungal pathogens) have been shown to attack planktonic blue-greens; certain chytrids specifically infest akinetes, other heterocysts. Bacterial pathogens of blue-greens seem all to belong to the group of Myxobacteriales. They can effect rapid lysis of a wide range of unicellular and filamentous blue-greens. Heterocysts and akinetes usually, though not invariably, remain unaffected. The bacteria attack always by means of end-on contact with the host cell; in this fashion they progress from cell to cell along a trichome.

Various viral pathogenes that infect blue-green cells have been isolated. All belong to the group of complex viruses or phages, and were named cyanophages. They exhibit some degree of host specifity. The LPP–1 virus, for example, is effective only against strains of *Lyngbya*, *Phormidium* and *Plectonema*. Phage C–1 lyses only *Cylindrospermum*, phage AR–1 attacks *Anabaenopsis*, while phages SM–1 and AS–1 are effective against the unicellular forms, *Synechococcus* and *Microcystis*. Viral numbers have been found to increase in lakes in response to the seasonal development of blue-green populations, and may have an important role in controlling them in nature.

The long-term approach to the control of blue-green growth in polluted waters is no doubt the systematic removal of major nutrients which are principally responsible for the extensive growth of planktonic populations. Waste-water purification is now increasingly practised in many countries, bringing about striking changes in water supplies.

7.4 The nutritive value of blue-greens

Blue-greens have recently been considered by several workers to be potentially excellent sources of food for animals and even for man. This may sound at first an astonishing idea but the introduction of blue-greens into the human diet would by no means constitute a genuine innovation.

In the northern parts of the African republic of Chad there are stretches of shallow alkaline lakes with high salt (sodium carbonate and bicarbonate) content which produce abundant growth of the planktonic *Spirulina platensis*. The planktonic material has been collected for food by the local people since ancient times and dried in the sun. The fried algal cake, called 'die', is being sold at the market of Fort Lamy and throughout the country. It is used for the preparation of soup and sauce, especially when meat is in short supply. Chemical analysis showed that the dry 'die' contains about 62% proteinaceous material, 16 to 18% carbohydrate, 2–3% lipid as well as mineral salts. It is likely that one of the basic food stuffs of the Aztecs was of a similar nature, which they collected from shallow Mexican lakes. Indeed, such lakes are currently commercially exploited for the mass production of *Spirulina*, and the blue-green crop is used as cattle fodder, for the extraction of various organic compounds, and more recently also for the production of nutritious 'health food'. *Nostoc* is utilized as food in Peru and in the Far East. *Microcystis* has been grown on domestic wastes and used to feed fish in lagoons and fish ponds in India and Pakistan.

Chemical analysis of the protein extracted from *Spirulina* has shown that it contains all the essential amino acids at a concentration which should make it suitable for human or animal consumption, with the exception of the relatively low quantities of sulphur-containing amino acids (cysteine and methionine). The nutritive value (protein index) of *Spirulina* was estimated to be round 50%. (For comparison, the protein index of groundnut is 54%, of soya meal 75%, beef 83% and egg 100%.) Feeding experiments with animals indicate that the *Spirulina* diet is acceptable and well digested by farm animals.

7.5 The use of blue-greens in agriculture

The Earth's human population is over 4 billions, and with the present rate of growth it may double by the end of this century. This situation will present an unprecedented demand for food in a world where malnutrition and starvation are already desperate problems. At the same time the prospects of increased food production are aggrevated by the ever increasing costs of fuel and chemical fertilizers. In view of the world's food crisis, attention has been turned lately to traditional farming practices and to more recent inventions which enable the more extensive and efficient utilization of biological nitrogen fertilizers.

Organic fertilizers are of great value for agriculture, not only in providing essential plant nutrients but for a number of other reasons. Natural manure, in contrast to chemical fertilizers, is a continuous source of nutrients, which are slowly released in accordance with requirements of plant growth. It is also a source of organic substrates which support the growth of indispensable microbial populations in the soil. Organic fertilizers greatly improve soil structure, prevent erosion and promote soil conservation. In addition they increase the efficient utilization of mineral fertilizers. Also, the use of biological fertilizers reduces the pollution arising from the excessive use and leaching of chemicals from arable land into natural waters and water reservoirs.

Many tropical paddy fields receive no chemical fertilizer nor natural manure, yet they remain productive and capable of supporting large populations with basic food. The fertility of paddy soil is maintained by the activities of heterocystous blue-greens which grow spontaneously and often luxuriantly in the waterlogged field. They provide fixed nitrogen to rice plants through both secretion of nitrogenous substances and on their decay and subsequent mineralization of organic substances in the soil.

In order to further their beneficial effects, blue-greens are grown on a large scale in India, the Philippines and China, and disseminated in paddy fields. The starter culture is a mixture of the most efficient N_2-fixing strains which grow well in the particular region. The cultures are used to inoculate outdoor nursery plots. The nursery cultures in turn provide the inoculum for the flooded rice fields, which are seeded at a concentration of about 750 kg per hectare. Within 10 to 15 days, growth of the blue-green populations in the paddy may reach 7.5 to 15 tonnes per hectare. It was estimated that in the Philippino rice fields blue-green algal growth adds up to 40 kg nitrogen per hectare per year to the soil. Field trails in India indicate that blue-green algal application results in about a 10 to 15% increase in crop yield under optimum conditions. In India blue-greens are also employed for the reclamation of alkaline and saline wastelands. The wasteland is enclosed by an earth embankment about 0.5 m high which promotes waterlogging during the rainy period. The rapid growth of blue-greens in the waterlogged plots decreases soil alkalinity and increases the nitrogen and organic content of the soil. In a few years time the wastelands can be converted to fertile fields.

The beneficial effect of N_2-fixation for crop production are even more apparent in paddy fields which are populated by the water fern *Azolla* during the waterlogged period (see § 5.4). The utility of *Azolla* as an efficient fertilizer of paddy fields has been known in North Vietnam for centuries. It is cultivated there over about 400 000 hectares of land as a specific crop for green compost and forage. The cultivation of *Azolla* has also been introduced in China, Thailand, Indoniesia and India to enahance rice production. Estimates of the annual quantities of N_2 fixed by *Azolla* range between 120 and 312 kg nitrogen per hectare. Trial applications on Californian rice fields have shown that the fern can accumulate over 60 kg nitrogen per hectare within 45 days. The plant supplies substantial amounts of geen manure (between 150–300 ton per hectare per year), which supports the growth of soil micro-organisms including heterotrophic N_2-fixers. The application of *Azolla* can increase rice yield by up to 38%; an equivalent increase would require the administration of 30 kg nitrogen per hectare in the form of chemical nitrogen fertilizer.

Many suggestions have been put forward recently for the application of blue-green innoculations to temperate cultivated areas, which are based on similar considerations to the practice of inoculating leguminous seeds with *Rhizobium* when a particular legume species is planted the first time in a field. Preliminary tests, however, suggest that the benefits of blue-green inoculations to the soil in terms of crop production are doubtful or insignificant. Some minor increase in crop yield occured even when dead cell material was distributed over the field, and was clearly due to the supply of organic matter to soil. The failure is attributed to several reasons. Inoculation must be applied early in spring before crop seeding in order to be effective, but soils are generally too cold then for the growth and N_2-fixation of blue-greens. Decay and mineralization of blue-green material is a much slower process in temperate soils, and therefore the N_2 fixed will not be available before the next growing season. The initial growth of blue-greens depends on soil moisture; dry soils may require irrigation. Also, present techniques for growing the inoculum are not economical. Above all the principal question is whether there is a need for inoculation. The presence of N_2-fixing blue-greens in neutral and slightly alkaline agricultural soil is well documented. (Long-term nitrogen balance studies on the famous Broadbalk plots at the Rothamsted Experimental Station, Hertfordshire, suggest that soil blue-greens, which often form a conspicuous cover over the soil, may fix N_2 at a rate of about 20 kg N ha^{-1} y^{-1} in plots to which no mineral or organic nitrogen fertilizers are added but other essential mineral nutrients are provided.) Thus it is obviously not the absence of blue-greens or the lack of suitable strains which must be corrected but the conditions which favour their growth and their N_2-fixing activity in temperate soil which require further attention and detailed investigation.

The development of blue-green populations in soils and paddy fields, as well as the successful application of *Azolla* nitrogen fertilizer greatly depend on a delicate biological balance in their environment. It was mentioned earlier that a variety of invertebrates commonly graze on blue-greens in rice fields (§ 7.3). Insect larvae preferentially feed on the leaves of *Azolla*. The multiplication of grazer populations may rapidly diminish the number of blue-greens and *Azolla* plants in paddies. The introduction of fish like *Tilapia mozambica*, which feed on algal grazers, appears to be an ideal choice to control the grazer population. Alternatively insecticides may have to be applied to protect the blue-green community or the *Azolla* crop. However, both blue-greens and *Azolla* have been shown to be susceptible to the effect of several pesticides used in common agricultural practice, although their sensitivity may vary considerably according to species and the chemical nature of the pesticides. In general blue-greens appear to be more resistant to pesticides than eukaryotic algae, and most strains tolerate the pesticide levels normally applied to agricultural land. The control of pests, as always, requires careful prior consideration of the effect of pesticides on the whole environment, and suggests the preferential use of highly selective pesticides and possibly the isolation of pesticide-resistant strains of blue-greens.

7.6 Solar energy conversion through biophotolysis

Heterocystous blue-greens possess the unique ability to simultaneously evolve O_2 in photosynthesis (in vegetative cells) and H_2 by nitrogenase catalyzed electron transfer to H^+-ions (in heterocysts), in the absence of N_2 or other substrates of nitrogenase (see § 3.7). This is the basis for the attempts of several workers to exploit this potential through the development of a 'biophotolytic system' for solar energy conversion. Following earlier laboratory studies successful outdoor systems were established in which *Anabaena cylindrica* was used to produce H_2 and O_2 in light when sparged continously with a gas mixture containing argon and CO_2. Thermodynamic efficiency of harvesting light energy and converting it into chemical energy through the combined action of the photosynthetic and nitrogenase systems ranges between 0.35 and 0.85%. This is a disappointingly low value, and clearly a great deal of improvement is necessary before such a system could become a feasible proposition for solar energy conversion.

We may nevertheless conclude that the utilization of blue-greens in food production and in solar energy conversion may hold immense potential for the future. There can be no doubt that the unique metabolic constitution of blue-greens could be exploited for man's economy. Progress in the study of the genetics of blue-greens may enable us to manipulate the N_2-fixation (*nif*) and associated genes, and produce strains which fix N_2, evolve H_2 or release ammonia with great efficiency.

Further Reading

ADAMS, D.G. and CARR, N.G. (1981). The developmental biology of heterocysts and akinete formation in cyanobacteria. *CRC Critical Review of Microbiology*, **9**, 45–100.

BOTHE, H. (1982). Hydrogen production by algae. *Experientia*, **38**, 53–64.

CARR, N.G. and WHITTON, B.A. (eds) (1982). *The Biology of Cyanobacteria*. Blackwell Scientific Publications, Oxford. 704 pp.

DESIKACHARY, T.V. (1959). *Cyanophyta*. Indian Council of Agricultural Research, New Delhi. 686 pp.

DOOLITTLE, W.F. (1979). The cyanobacterial genome, its expression, and the control of that expression. In *Advances in Microbial Physiology*, **20**, 2–102. Academic Press, London.

FAY, P. (1980). Nitrogen fixation in heterocysts. In *Recent Advances in Biological Nitrogen Fixation*, Subba-Rao, N.S. (ed.), pp. 121–65. Edward Arnold, London.

FOGG, G.E., STEWART, W.D.P., FAY, P. and WALSBY, A.E. (1973) *The Blue-Green Algae*. Academic Press, London. 459 pp.

FRITSCH, F.E. (1945). *The Structure and Reproduction of Algae*. Vol. 2. University Press, Cambridge. 939 pp.

GLAZER, A.N. (1982). Phycobilisomes: structure and dynamics. *Annual Review of Microbiology*, **36**, 173–98.

PADAN, E. (1979). Facultative anoxygenic photosynthesis in cyanobacteria. *Annual Review of Plant Physiology*, **30**, 27–40.

REYNOLDS, C.S. and WALSBY, A.E. (1975). Water blooms. *Biological Reviews*, **50**, 437–81.

RIPPKA, R., DERUELLES, J., WATERBURY, J.B., HERDMAN, M. and STANIER, R.Y. (1979). Generic assignments, strain histories and properties of pure cultures of cyanobacteria. *Journal of general Microbiology*, **111**, 1–61.

SCHOPF, J.W. (1970). Precambrian micro-organisms and evolutionary events prior to the origin of vascular plants. *Biological Reviews*, **45**, 319–52.

SHELEF, G. and SOEDER, C.J. (eds) (1980). *Algae Biomass Production and Use*. Elsevier/North Holland Publications, Amsterdam. 825 pp.

STANIER, R.Y. and COHEN-BAZIRE, G. (1979). Phototrophic prokaryotes: the cyanobacteria. *Annual Review of Microbiology*, **31**, 225–74.

STEWART, W.D.P. (1978). Nitrogen-fixing cyanobacteria and their associations with eukaryotic plants. *Endeavour*, **2**, 170–79.

STEWART, W.D.P. (1980). Some aspects of structure and function in N_2-fixing cyanobacteria. *Annual Review of Microbiology*, **34**, 497–536.

WOLK, C.P. (1973). Physiology and cytological chemistry of blue-green algae. *Bacteriological Reviews*, **37**, 32–101.

Index